D1187463

Reversals of the Earth's magnetic field

To Ann

Reversals of the Earth's magnetic field

J A Jacobs

Department of Earth Sciences, University of Cambridge

Adam Hilger Ltd, Bristol

British Library Cataloguing in Publication Data

Jacobs, J. A.
   Reversals of the earth's magnetic field
   1. Magnetism, Terrestrial
   I. Title
   538'.7   QC815.2

   ISBN 0-85274-442-0

89 – 1412

Published by Adam Hilger Ltd, Techno House, Redcliffe Way,
Bristol BS1 6NX

Printed in Great Britain by J W Arrowsmith Ltd, Bristol

Contents

Preface

In June 1967, Sir Edward Bullard delivered The Bakerian Lecture to The Royal Society. The title of his talk was 'Reversals of the Earth's magnetic field'. Such reversals have played a key role in the development of plate tectonics which has revolutionised geological thinking about the evolution of the outer surface of our planet. Yet, in spite of a great increase in the amount of data accumulated, the physics of the reversal mechanism is still not understood— nor, for that matter, is the detailed mechanism of the generation of the field itself. This book, which was essentially completed by the end of 1982, is an attempt to bring together this large body of additional data (chapter 3) and some of the more recent ideas on possible reversal mechanisms (chapter 5). The case for self-reversal is discussed in chapter 2. Particular attention is given to excursions of the geomagnetic field (or aborted reversals?), which are discussed in chapter 4.

Reversals have also contributed much to the chronology of the last few hundred million years. This subject (magnetostratigraphy) is discussed in chapter 6. Space prevents a detailed account of the vast literature on this subject. There have also been many attempts to correlate changes in the Earth's magnetic field with climate, and the literature here is even greater. Chapter 7 discusses some of these questions—particularly that of possible correlations between reversals of the Earth's magnetic field, changing climatic conditions and ice ages. Here one must be wary of statistics based on very limited time series and the popular appeal of relating reversals to many other geophysical phenomena.

I wish to acknowledge the award of an Emeritus Fellowship from The Leverhulme Trust which enabled me to complete this monograph.

Acknowledgments

The author and publisher are grateful to the following publishers for permission to reproduce the figures listed:

American Association for the Advancement of Science: figures 3.1, 3.2, 3.3, 7.1, 7.2, 7.3

American Geophysical Union: figures 3.15, 3.16, 3.18, 3.19, 3.20, 3.21, 4.7, 5.9, 5.11, 5.17, 6.1, 6.6, 6.7, 6.8, 7.5, 7.9, 7.11

American Institute of Physics: figures 5.1, 5.2

Blackwell Scientific Publications Ltd: figures 2.4, 3.5, 3.6, 3.7, 3.8, 3.9, 3.10, 3.13, 3.22, 3.23, 5.19

Cambridge University Press: figures 5.3, 5.4, 5.5

Carnegie Institute: figures 1.5, 1.6

Elsevier Scientific Publishing Co: figures 3.4, 3.12, 3.14, 3.25, 4.1, 4.2, 4.3, 4.6, 5.6, 5.7, 5.8, 5.10, 5.15, 5.16, 5.22, 7.4

Geological Society of America: figures 2.1, 6.3, 6.4, 6.5, 7.10, 7.12

John Wiley and Sons Ltd: figure 3.17

Macmillan Journals Ltd: figures 3.27, 4.4, 4.5, 6.2, 7.7, 7.8

Scientific American Inc: figure 7.13

Society of Terrestrial Magnetism and Electricity of Japan: figures 1.2, 1.3, 3.24, 5.20, 5.21, 7.6

The Royal Society: figures 3.11, 3.26

University of Chicago Press: figures 5.12, 5.13, 5.14

University of Pittsburgh: figure 5.18

US Department of Commerce: figure 1.4

1 The Earth's magnetic field

1.1 Introduction

In geomagnetism we are measuring extremely small magnetic fields—at its strongest near the poles, the Earth's magnetic field is several hundred times weaker than that between the poles of a toy horseshoe magnet. In a magnetic compass, the needle is weighted so that it will swing in a horizontal plane, its deviation from geographical north being called the declination, D. A non-magnetic needle which is balanced horizontally on a pivot becomes inclined to the vertical when magnetised. Over most of the northern hemisphere the north-seeking end of the needle will dip downwards, the angle it makes with the horizontal being called the magnetic dip or inclination, I. The total intensity F, the declination D and the inclination I completely define the magnetic field at any point. The horizontal and vertical components of F are denoted by H and Z. H may be further resolved into two components X and Y, X being the component along the geographical meridian (northward) and Y the orthogonal component (eastward). Figure 1.1 illustrates these different magnetic elements. They are simply related to one another by the following equations

$$H = F \cos I \qquad Z = F \sin I \qquad \tan I = Z/H \qquad (1.1)$$

$$X = H \cos D \qquad Y = H \sin D \qquad \tan D = Y/X \qquad (1.2)$$

$$F^2 = H^2 + Z^2 = X^2 + Y^2 + Z^2. \qquad (1.3)$$

The variation of the magnetic field over the Earth's surface is best illustrated by isomagnetic charts, i.e. maps on which lines are drawn through points at which a given magnetic element has the same value. Contours of equal intensity in any of the elements X, Y, Z, H or F are called isodynamics. Figure 1.2 and figure 1.3 are world maps showing contours of equal declination (isogonics) and equal inclination (isoclinics) for the year 1980. Palaeomagnetists have traditionally used the oersted as the unit of magnetic field strength and the gauss (Γ) as the unit of magnetic induction. The distinction is

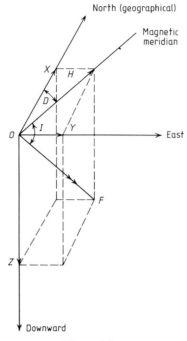

Figure 1.1

somewhat pedantic in geophysical applications since the permeability of air is virtually unity in cgs units. In SI units which will be used throughout this book,

$$1 \; \Gamma = 10^{-4} \; \text{Wb m}^{-2} \; (\text{weber/m}^2) = 10^{-4} \; \text{T (tesla)}.$$

Since in geomagnetism we are measuring extremely small magnetic fields, a more convenient unit is the gamma (γ) defined as

$$1 \; \gamma = 10^{-9} \; \text{T} = 1 \; \text{nT}.$$

Apart from its spatial variation, the Earth's magnetic field also shows temporal changes ranging from variations on a timescale of seconds to secular variations on a timescale of hundreds of years and on an even longer timescale to complete reversals of polarity. The short period, transient, variations are due to external influences and have no lasting effect on the Earth's main magnetic field which is of internal origin. They will not be discussed at all in this book. Variations over $10-10^4$ yr may be determined from archaeomagnetic and palaeomagnetic studies of the secular variation. This time range is probably characteristic of core fluid motions. If successive annual mean values of a magnetic element are obtained from a particular station, it is found that these secular changes are in the same sense over a long period of time, although the rate of change does not usually remain constant. Figure 1.4 shows the

IGRF 1980   Declination (*D*)   (degrees)

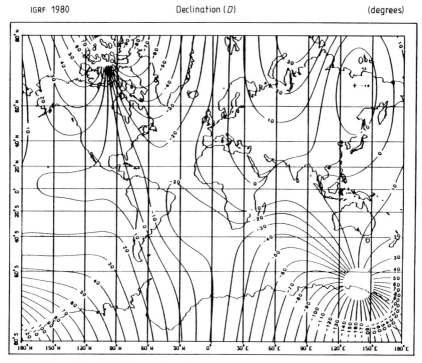

Figure 1.2 World map showing contours of equal declination *D* for 1980 (after Peddie 1982).

changes in declination and inclination at London, Boston and Baltimore. The declination at London was $11\frac{1}{2}°$E in 1580 and $24\frac{1}{2}°$W in 1819, a change of almost 36° in 240 years. Lines of equal secular change (isopors) in an element form sets of ovals centring on points of local maximum change (isoporic foci). Figures 1.5 and 1.6 show the secular change in *Z* for the years 1922.5 and 1942.5. It can be seen that the secular variation is a regional rather than a planetary phenomenon and that considerable changes can take place in the general distribution of isopors even within 20 years. The secular variation is anomalously large and complicated over and around Antarctica; on the other hand it is markedly smaller in the Pacific hemisphere (between about 120°E and 80°W). There is a strong secular change focus in the Atlantic where the vertical intensity is changing non-linearly. *Z* was approximately $-50 \, \text{nT yr}^{-1}$ in the 1960s and $-150 \, \text{nT yr}^{-1}$ in 1978. The existence of the Pacific (magnetically) quiet zone and other features of the secular variation suggest that generation of the Earth's magnetic field may be partly controlled by lateral inhomogeneities at the mantle-core boundary (MCB). Reversals occur at widely varying intervals from about 30 000 yr to more than 10 Myr and are more likely to be due to random instabilities in the core fluid. Finally, variations in the frequency of reversals which have time constants of the order

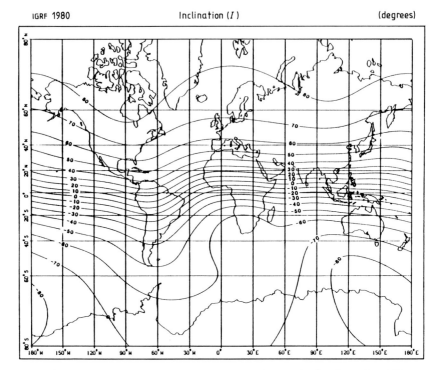

Figure 1.3   World map showing contours of equal inclination *I* for 1980 (after Peddie 1982).

of 50 Myr may represent changing conditions at the MCB. These questions will be discussed in later chapters.

In 1839 Gauss showed that the field of a uniformly magnetised sphere, which is the same as that of a dipole at its centre, is an excellent first approximation to the Earth's magnetic field. Gauss further analysed the irregular part of the Earth's field, i.e. the difference between the actual observed field and that due to a uniformly magnetised sphere, and showed that both the regular and irregular components of the Earth's field are of internal origin.

Since the north-seeking end of a compass needle is attracted towards the northern regions of the Earth, those regions must have opposite polarity. Consider therefore the field of a uniformly magnetised sphere whose magnetic axis runs north–south, and let *P* be any external point distant *r* from the centre *O* and θ the angle *NOP*, i.e. θ is the magnetic co-latitude (see figure 1.7). If *m* is the magnetic moment of a geocentric dipole directed along the axis, the potential at *P* is

$$V = \frac{m}{4\pi} \frac{\cos \theta}{r^2}.$$
(1.4)

The inward radial component of force corresponding to the magnetic

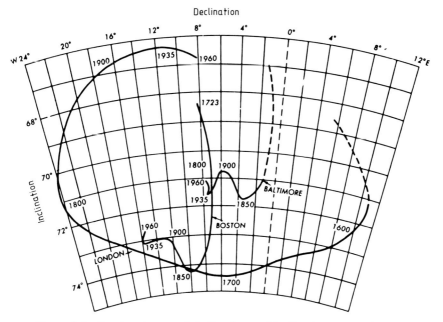

Figure 1.4 Secular change of declination and inclination at London, Boston and Baltimore (after Nelson *et al* 1962).

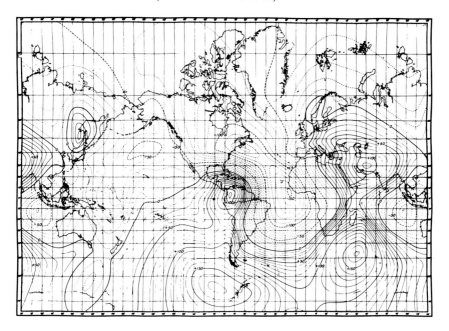

Figure 1.5 World map showing the geomagnetic secular variation of the vertical component Z. Epoch 1922.5 (after Vestine *et al* 1947).

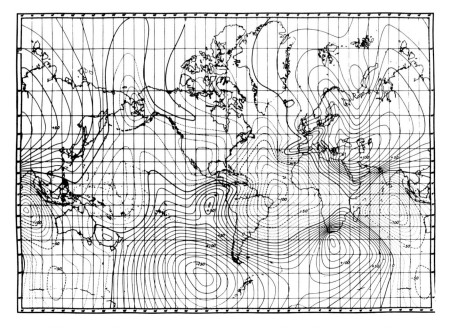

Figure 1.6   World map showing the geomagnetic secular variation of the vertical component Z. Epoch 1942.5 (after Vestine *et al* 1947).

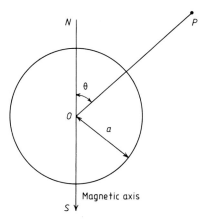

Figure 1.7

component Z is given by

$$Z = -\mu_0 \frac{\partial V}{\partial r} = \frac{\mu_0 m}{2\pi} \frac{\cos \theta}{r^3} \tag{1.5}$$

and the component at right angles to OP in the direction of decreasing θ, corresponding to the magnetic component H, by

$$H = -\mu_0 \frac{1}{r}\frac{\partial V}{\partial \theta} = \frac{\mu_0 m \sin \theta}{4\pi} \frac{1}{r^3} \qquad (1.6)$$

where μ_0 is the permeability of free space.

The inclination I is then given by

$$\tan I = Z/H = 2 \cot \theta \qquad (1.7)$$

and the magnitude of the total force F by

$$F = (H^2 + Z^2)^{1/2} = \frac{\mu_0 m}{4\pi r^3}(1 + 3 \cos^2 \theta)^{1/2}. \qquad (1.8)$$

Thus intensity measurements are a function of latitude.

The geomagnetic poles, i.e. the points where the axis of the geocentric dipole which best approximates the Earth's field meets the surface of the Earth, are situated at approximately 79°N, 70°W and 79°S, 110°E. The geomagnetic axis is thus inclined at about 11° to the Earth's geographical axis.

One of the most interesting results of palaeomagnetic studies is that many igneous rocks show a permanent magnetisation approximately opposite in direction to that of the present field. Reverse magnetisation was first discovered in 1906 by Brunhes in a lava from the Massif Central mountain range in France—since then examples have been found in almost every part of the world. About one-half of all rocks measured are found to be normally magnetised and one-half reversely. Dagley *et al* (1967) carried out an extensive palaeomagnetic survey of Eastern Iceland sampling some 900 separate lava flows lying on top of each other. The direction of magnetisation of more than 2000 samples representative of individual lava flows was determined covering a time interval of 20 Myr. At least 61 polarity zones, or 60 complete changes of polarity were found giving an average rate of at least 3 inversions/Myr. The same pattern of reversals observed in igneous rocks has also been found in deep-sea sediments (see e.g. Opdyke *et al* 1966).

1.2 Spherical harmonic analysis of the Earth's magnetic field

Assuming that there is no magnetic material near the ground, the Earth's magnetic field can be derived from a potential function V which satisfies Laplace's equation and can thus be represented as a series of spherical harmonics

$$V = \frac{a}{\mu_0} \sum_{n=1}^{\infty} \sum_{m=0}^{n} P_n^m(\cos \theta) \left\{ \left[c_n^m \left(\frac{r}{a}\right)^n + (1 - c_n^m)\left(\frac{a}{r}\right)^{n+1} \right] A_n^m \cos m\phi \right.$$
$$\left. + \left[s_n^m \left(\frac{r}{a}\right)^n + (1 - s_n^m)\left(\frac{a}{r}\right)^{n+1} \right] B_n^m \sin m\phi \right\}. \qquad (1.9)$$

Written in this form the coefficients A_n^m and B_n^m have the dimensions of magnetic field, c_n^m and s_n^m are numbers lying between 0 and 1, and represent the fractions of the harmonic terms $P_n^m(\cos\theta)\cos m\phi$ and $P_n^m(\cos\theta)\sin m\phi$ in the expansion of V which, on the surface of the sphere $(r \sim a)$, are due to matter outside the sphere. There is no term with $n=0$, which would correspond to a magnetic monopole within the Earth. It is also assumed that there are no electric currents flowing across the surface of the Earth—if there were they would set up a non-potential field and thus contribute a part of the Earth's magnetic field which could not be represented by equation (1.9).

The potential V cannot be measured directly; what can be determined are the three components of force $X=(\mu_0/r)/(\partial V/\partial\theta)$ (horizontal, northward), $Y=(-\mu_0/r\sin\theta)(\partial V/\partial\phi)$ (horizontal, eastward) and $Z=\mu_0(\partial V/\partial r)$ (vertical, downward) at the Earth's surface, $r=a$. Z (at $r=a$) may be expanded as a series of spherical harmonics

$$Z=\mu_0\frac{\partial V}{\partial r}=\sum_{n=1}^{\infty}\sum_{m=0}^{n}P_n^m(\cos\theta)(\alpha_n^m\cos m\phi+\beta_n^m\sin m\phi) \qquad (1.10)$$

and the coefficients α_n^m, β_n^m determined from the observed values of Z.

By differentiating equation (1.9) with respect to r and then writing $r=a$, we have

$$Z=\mu_0\frac{\partial V}{\partial r}=\sum_{n=1}^{\infty}\sum_{m=0}^{n}P_n^m(\cos\theta)\{[nc_n^m-(n+1)(1-c_n^m)]A_n^m\cos m\phi$$

$$+[ns_n^m-(n+1)(1-s_n^m)]B_n^m\sin m\phi\}. \qquad (1.11)$$

The coefficients of each separate harmonic term for each n and m must be equal in the two expansions of Z given by equations (1.10) and (1.11). Hence

$$\alpha_n^m=[nc_n^m-(n+1)(1-c_n^m)]A_n^m$$
$$\beta_n^m=[ns_n^m-(n+1)(1-s_n^m)]B_n^m. \qquad (1.12)$$

Again from an analysis of the observed values of X and Y, the coefficients in the following two expansions derived from equation (1.9) may be obtained

$$Y_{r=a}=\left(\frac{-\mu_0}{r\sin\theta}\frac{\partial V}{\partial\phi}\right)_{r=a}$$

$$=\frac{1}{\sin\theta}\sum_{n=1}^{\infty}\sum_{m=0}^{n}P_n^m(\cos\theta)(mA_n^m\sin m\phi-mB_n^m\cos m\phi) \qquad (1.13)$$

$$X_{r=a}=\left(\frac{\mu_0}{r}\frac{\partial V}{\partial\theta}\right)_{r=a}$$

$$=\sum_{n=1}^{\infty}\sum_{m=0}^{n}\frac{d}{d\theta}P_n^m(\cos\theta)(A_n^m\cos m\phi+B_n^m\sin m\phi). \qquad (1.14)$$

Both these equations contain A_n^m and B_n^m and if values of X are known all over the world, values of Y can be deduced. If there is disagreement between observed and calculated values of Y, it would imply that the field was not completely derivable from a potential V and hence that Earth–air currents do exist. When Gauss first carried out such calculations in 1839 he found no discrepancy. From a knowledge of the coefficients A_n^m, B_n^m, α_n^m and β_n^m, equations (1.12) determined c_n^m and s_n^m. Gauss found from data available at that time that $c_n^m = s_n^m = 0$, i.e. the source of the Earth's magnetic field is entirely internal. The coefficients of the field of internal origin are

$$g_n^m = (1 - c_n^m)A_n^m \qquad h_n^m = (1 - s_n^m)B_n^m \qquad (1.15)$$

and are known as Gauss coefficients. If the external field is negligible, equations (1.15) reduce to $g_n^m = A_n^m$, and $h_n^m = B_n^m$. Values of these coefficients as obtained by different investigators since the time of Gauss are given in table 1.1. It is clear that by far the most important contribution to V comes from the term containing g_1^0 which is proportional to $P_1(\cos \theta)/r^2$, i.e. $\cos \theta/r^2$, and corresponds to the field of a geocentric dipole oriented along the z axis. Harmonic analysis of most force fields yields a first degree term, but it must be emphasised that the formal calculation of the term provides little insight into whether it describes a distinct physical entity or process. The g_1^1 term is proportional to $\sin \theta \cos \phi/r^2$. If γ is the angle between the x and r axes (see figure 1.8),

$$\cos \gamma = \sin \theta \cos \phi \qquad (1.16)$$

so that the g_1^1 term corresponds to a geocentric dipole oriented in the x direction. Similarly the h_1^1 term is proportional to $\sin \theta \sin \phi/r^2$ and corresponds to a geocentric dipole oriented in the y direction. The terms g_1^0, g_1^1, h_1^1 can be combined to give the total geocentric dipole which at present is inclined at about $11\frac{1}{2}°$ to the rotation axis. The terms involving $n = 2$ (r^{-3} in the potential) represent a geocentric quadrupole, the terms $n = 3$ (r^{-4} in the potential) a geocentric octupole etc. It must be emphasised that a spherical harmonic analysis is nothing more than a mathematical procedure for representing the field by hypothetical sources at the Earth's centre. The actual sources of the Earth's magnetic field must almost certainly lie in the fluid outer core and not in the solid inner core.

It is sometimes convenient to represent the Earth's field as a dipole displaced from the centre a distance d along the z axis. The potential of such an offset dipole is proportional to

$$\cos \theta(r^2 + d^2 - 2rd \cos \theta)^{-1} = \frac{\cos \theta}{r^2}\left(1 - \frac{2d}{r}\cos \theta + \frac{d^2}{r^2}\right)^{-1}. \qquad (1.17)$$

Expanding in a Taylor series for $d < r$ it can be seen that the offset dipole is equivalent to a dipole plus a series of higher degree multipole fields at the Earth's centre. The larger the value of d, the larger the number of zonal

Table 1.1 Dipole and quadrupole Gauss coefficients (10^{-4}T) and polar angles (degrees) at various epochs. (After McDonald and Gunst 1967.)

Author	Epoch	g_1^0	g_1^1	h_1^1	g_2^0	g_2^1	g_2^2	h_2^1	h_2^2	θ polar (deg)
Erman–Petersen	1829	−0.3201	−0.0284	0.0601	−0.0008	0.0257	−0.0014	−0.0004	0.0146	11.7
Gauss	1835	−0.3235	−0.0311	0.0625	0.0051	0.0292	−0.0002	0.0012	0.0157	12.2
Adams	1845	−0.3219	−0.0278	0.0578	0.0009	0.0284	0.0004	−0.0010	0.0135	11.3
Adams	1880	−0.3168	−0.0243	0.0603	−0.0049	0.0297	−0.0006	−0.0075	0.0149	11.6
Neumayer	1880	−0.3157	−0.0248	0.0603	−0.0053	0.0288	0.0065	−0.0075	0.0146	11.7
Fritsche	1885	−0.3164	−0.0241	0.0591	−0.0035	0.0286	0.0068	−0.0075	0.0142	11.4
Schmidt	1885	−0.3174	−0.0236	0.0598	−0.0050	0.0278	0.0065	−0.0071	0.0149	11.5
Dyson–Furner	1922	−0.3095	−0.0226	0.0592	−0.00887	0.02991	0.01443	−0.01241	0.00843	11.4
Jones–Melotte	1942.5	−0.3039	−0.0218	0.0555	−0.0117	0.02940	0.0156	−0.0150	0.0051	11.1
Vestine et al	1905	−0.31423	−0.02270	0.05981	−0.00773	0.02952	0.01107	−0.01051	0.01156	11.4
Vestine et al	1915	−0.31176	−0.02176	0.05912	−0.00842	0.02940	0.01345	−0.01144	0.00986	11.4
Vestine et al	1925	−0.30892	−0.02166	0.05839	−0.00946	0.02946	0.01510	−0.01284	0.00814	11.4
Vestine et al	1935	−0.30662	−0.02129	0.05792	−0.01086	0.02959	0.01608	−0.01460	0.00676	11.4
Vestine et al	1945	−0.30570	−0.02116	0.05805	−0.01265	0.02960	0.01632	−0.01658	0.00535	11.4
Afanasieva	1945	−0.3032	−0.0229	0.0590	−0.0125	0.0288	0.0150	−0.0146	0.0048	11.8
U.S.C. & G.S.	1945	−0.3057	−0.0219	0.0579						11.4
Franselau–Kautzleben	1945	−0.30668	−0.02160	0.0577	−0.01279	0.029596	0.01547	−0.01673	0.005811	11.4
U.S.C. & G.S.	1955	−0.3046	−0.0212	0.0576						11.4
Finch–Leaton	1955	−0.3055	−0.0227	0.0590	−0.0152	0.0303	0.0158	−0.1090	0.0024	11.7
Nagata–Oguti	1958.5	−0.3045	−0.0222	0.0584	−0.0151	0.0295	0.0149	−0.0194	0.0021	11.6
Cain et al	1959	−0.30674	−0.01923	0.05762	−0.02055	0.0344	0.0124	−0.01977	0.00671	11.2
Fougere	1960	−0.30509	−0.02181	0.05841	−0.01464	0.02971	0.01673	−0.01988	0.00198	11.6
Adam et al	1960	−0.3046	−0.0214	0.0580	−0.0150	0.0299	0.0164	−0.0194	0.0027	11.5
Jensen–Cain	1960	−0.30411	−0.02147	0.05799	−0.01602	0.02959	0.01545	−0.01912	0.00812	11.5
Leaton et al	1965	−0.30375	−0.02087	0.05769	−0.01648	0.02954	0.01579	−0.01995	0.00116	11.4
Hurwitz et al	1965	−0.30388	−0.02117	0.05760	−0.01640	0.02983	0.01583	−0.02004	0.00125	11.4

mean 11.49

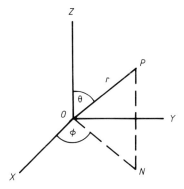

Figure 1.8

harmonics required to approximate the offset dipole. Bartels (1936) showed
that

$$g_2^0/2g_1^0 = d/a \qquad \text{and} \qquad g_3^0/g_2^0 = 3d/2a. \tag{1.18}$$

This is discussed further in §5.6. The above example illustrates the non-
uniqueness of a spherical harmonic analysis and emphasises that separate
terms in such an analysis do not represent separate real magnetic sources. An
outstanding feature of table 1.1 is the secular variation of the individual
coefficients which is clearly apparent in spite of individual scatter. There
appears to have been an overall decrease in the dipole moment of the Earth's
field of about 5 % per century whilst the quadrupole moment has increased by
approximately 40% per century since AD 1800. Over the same time interval,
the inclination of the dipole axis appears to have remained sensibly constant.

  The non-dipole components of the Earth's field, though much weaker than
the dipole component, show more rapid changes. The timescale of the non-
dipole changes is measured in decades and that of the dipole in centuries. The
isoporic foci also drift westward at a fraction of a degree per year. Both
eastward and westward drifts of the non-dipole field have been inferred from
palaeomagnetic studies. Observatory records from Sitka, Alaska, indicate an
eastward drift during the past 60 years, in contrast to the predominance of a
westward direction of drift observed over most areas of the world for the past
several hundred years (Skiles 1970, Yukutake 1962).

  Yukutake and Tachinaka (1969) have shown that the non-dipole field can
be decomposed into two parts, one standing and the other drifting westward at
about the same rate as the secular variation. The standing and drifting parts
are approximately of the same size and intensity. The drifting field consists
mainly of low harmonics ($n \leqslant 3$), whereas the standing field has a more
complicated distribution. Explanations for the non-dipole field are of two
types: one postulates eddy currents in the outer core interacting with the
toroidal field (e.g. Bullard *et al* 1950), the other suggests that magnetohydrody-

namic waves in the outer core propagate eastward (with periods of the order of days) or westward (with periods of tens of years or more) (Hide 1966, Skiles 1972a, b). Either type of mechanism requires some method of exciting the disturbance, and bumps on the MCB have been suggested (Hide 1966, Hide and Malin 1970, Moffatt and Dillon 1976). Additional reasons for such core–mantle topography will be discussed later.

A number of people have attempted to carry out spherical harmonic analyses of the palaeomagnetic field of the Earth back through the Mesozoic and into the Palaeozoic. It is not surprising that different conclusions have been reached since the past distribution of the continents plays a critical role in any such spherical harmonic analysis, and the distribution of continents through the late Palaeozoic is not known for certain. A meaningful global analysis of the palaeomagnetic field is impossible when the observation points cover only about one-third of the Earth's surface and their relative positions are not precisely known. Although there is still a poor distribution and uneven quality of the raw data from ground studies, the situation has improved considerably through the use of aircraft and more recently with satellites. It has become apparent that secular changes cannot be extrapolated accurately over intervals longer than about four or five years.

In practice the series expansion of the magnetic potential V is truncated at a maximum value N of m and n, usually in the range 8–15. The International Association of Geomagnetism and Aeronomy (IAGA) adopted in 1968 an International Geomagnetic Reference Field (IGRF) describing the main geomagnetic field at 1965 by means of 80 spherical harmonic coefficients ($N = 8$). An additional set of 80 coefficients describing the secular variation was included for use in extending the main-field model in time, both backward (not earlier than 1955) and forward (not later than 1975). By the early 1970s it was becoming obvious that inaccuracies in the secular variation coefficients were causing unacceptably large errors in field values computed for current epochs from the model and the IGRF was revised by IAGA in 1975. This revision was limited to the provision of a revised set of 80 secular variation coefficients to be used for deriving field values for dates after 1975, the original and revised versions of the IGRF being continuous at 1975. By the late 1970s the cumulative effects of uncertainties in the secular variation models led to unacceptable inaccuracies in the IGRF and a second revision was made by IAGA in 1981. By that time the results of NASA's *Magsat* satellite mission had been analysed. *Magsat* was launched on 30 October 1979 into a sun-synchronous (twilight) orbit with an apogee of 561 km and a perigee of 352 km. Re-entry occurred on 11 June 1980. The instrumentation included a caesium vapour magnetometer to measure the magnitude of the field, a fluxgate magnetometer to measure the field components and an optical system to measure the orientation of the fluxgate magnetometer. The scalar magneto-meter was used to calibrate the vector magnetometer in flight. The data are estimated to be accurate to about 2 nT in magnitude and 6 nT in each component. Preliminary results of the *Magsat* mission have been given in *Geophysical Research Letters* (Vol 9, No 4, 1982).

The new version of the IGRF consists of five models: four models of the main geomagnetic field for 1965, 1970, 1975 and 1980 and a model of the secular variation valid for the interval 1980–1985. The four main-field models each consist of 120 coefficients ($N = 10$) whilst the secular variation model has, as before, 80 coefficients. Values of the coefficients are given in table 1.2. The main field models for 1965, 1970 and 1975 are designated Definitive International Geomagnetic Reference Fields (DGRF 1965, DGRF 1970 and DGRF 1975, respectively) since further revision is not envisaged. Linear interpolation between neighbouring models is to be used for dates that lie between the epochs of the models. The main-field model for 1980 is not continuous with the earlier series of IGRF models. For dates between 1975 and 1980 a Provisional International Geomagnetic Reference Field (PGRF 1975) is defined by linear interpolation between DGRF 1975 and the main-field coefficients of IGRF (1980). The models were derived from three sets of proposed models by taking weighted means, the weights being assigned according to the apparent accuracy of the proposed models. A special issue of the *Journal of Geomagnetism and Geoelectricity* (Vol 34, No 6, 1982) is devoted to papers on the revision of the IGRF (Peddie 1982).

1.3 Origin of the Earth's magnetic field

There has been much speculation on the origin of the Earth's magnetic field and its secular variation and many possible sources have been suggested, most of which have proved to be inadequate. The only possible means seems to be some form of electromagnetic induction, electric currents flowing in the Earth's fluid, electrically conducting core. Palaeomagnetic measurements have shown that the Earth's main field has existed throughout geologic time and that its strength has never differed significantly from its present value. In a bounded, stationary, electrically conducting body, any system of electric currents will decay. The field or current may be analysed into normal modes each of which decays exponentially with its own time constant. The time constant is proportional to σl^2 where σ is the electrical conductivity and l a characteristic length representing the distance in which the field changes by an appreciable amount. For a sphere the size of the Earth the most slowly decaying mode is reduced to $1/e$ of its initial strength in a time of the order of 100 000 yr. Since the age of the Earth is more than 4000 Myr the geomagnetic field cannot be a relic of the past, and a mechanism must be found for generating and maintaining electric currents to sustain the field. The most likely source of the electromotive force needed to maintain these currents is the motion of core material across the geomagnetic lines of force. The study of this process, in which the currents generated reinforce the magnetic field which

Table 1.2 Spherical harmonic coefficients of the International Geomagnetic Reference Field 1980.

	n	m	DGRF 1965 nT	1970 nT	1975 nT	IGRF 1980 nT	1980–85 nT yr⁻¹		n	m	DGRF 1965 nT	1970 nT	1975 nT	IGRF 1980 nT	1980–85 nT yr⁻¹
g	1	0	−30334	−30220	−30100	−29988	22.4	h	7	6	−23	−23	−23	−23	−0.1
g	1	1	−2119	−2068	−2013	−1957	11.3	g	7	7	1	−2	−5	−2	0.0
h	1	1	5776	5737	5675	5606	−15.9	h	7	7	−12	−11	−12	−10	1.1
g	2	0	−1662	−1781	−1902	−1997	−18.3	g	8	0	13	14	14	20	0.8
g	2	1	2997	3000	3010	3028	3.2	g	8	1	5	6	6	7	−0.2
h	2	1	−2016	−2047	−2067	−2129	−12.7	h	8	1	7	7	6	7	−0.1
g	2	2	1594	1611	1632	1662	7.0	g	8	2	−4	−2	−1	1	−0.3
h	2	2	114	25	−68	−199	−25.2	h	8	2	−12	−15	−16	−18	−0.7
g	3	0	1297	1287	1276	1279	0.0	g	8	3	−14	−13	−12	−11	0.3
g	3	1	−2038	−2091	−2144	−2181	−6.5	h	8	3	9	6	4	4	0.0
h	3	1	−404	−366	−333	−335	0.2	g	8	4	0	−3	−8	−7	−0.8
g	3	2	1292	1278	1260	1251	−0.7	h	8	4	−16	−17	−19	−22	−0.8
h	3	2	240	251	262	271	2.7	g	8	5	8	5	4	4	−0.2
g	3	3	856	838	830	833	1.0	h	8	5	4	6	6	9	0.2
h	3	3	−165	−196	−223	−252	−7.9	g	8	6	−1	0	0	3	0.7
g	4	0	957	952	946	938	−1.4	h	8	6	24	21	18	16	0.2
g	4	1	804	800	791	783	−1.4	g	8	7	11	11	10	7	−0.3
h	4	1	148	167	191	212	4.6	h	8	7	−3	−6	−10	−13	−1.1
g	4	2	479	461	438	398	−8.2	g	8	8	4	3	1	−1	1.2
h	4	2	−269	−266	−265	−257	1.6	h	8	8	−17	−16	−17	−15	0.8
g	4	3	−390	−395	−405	−419	−1.8	g	9	0	8	8	7	6	
h	4	3	13	26	39	53	2.9	g	9	1	10	10	10	11	
g	4	4	252	234	216	199	−5.0	h	9	1	−22	−21	−21	−21	
h	4	4	−269	−279	−288	−298	0.4	g	9	2	2	2	2	2	
g	5	0	−219	−216	−218	−219	1.5	h	9	2	15	16	16	16	
g	5	1	358	359	356	357	0.4	g	9	3	−13	−12	−12	−12	
h	5	1	19	26	31	46	1.8	h	9	3	7	6	7	9	
g	5	2	254	262	264	261	−0.8	g	9	4	10	10	10	9	
h	5	2	128	139	148	149	−0.4	h	9	4	−4	−4	−4	−5	
g	5	3	−31	−42	−59	−74	−3.3	g	9	5	−1	−1	−1	−3	
h	5	3	−126	−139	−152	−150	0.0	h	9	5	−5	−5	−5	−7	
g	5	4	−157	−160	−159	−162	0.2	g	9	6	−1	0	−1	−1	
h	5	4	−97	−91	−83	−78	1.3	h	9	6	10	10	10	9	
g	5	5	−62	−56	−49	−48	1.4	g	9	7	5	3	4	7	
h	5	5	81	83	88	92	2.1	h	9	7	10	11	11	10	
g	6	0	45	43	45	49	0.4	g	9	8	1	1	1	1	
g	6	1	61	64	66	65	0.0	h	9	8	−4	−2	−3	−6	
h	6	1	−11	−12	−13	−15	−0.5	g	9	9	−2	−1	−2	−5	
g	6	2	8	15	28	42	3.4	h	9	9	1	1	1	2	
h	6	2	100	100	99	93	−1.4	g	10	0	−2	−3	−3	−3	
g	6	3	−228	−212	−198	−192	0.8	g	10	1	−3	−3	−3	−4	
h	6	3	68	72	75	71	0.0	h	10	1	2	1	1	1	
g	6	4	4	2	1	4	0.8	g	10	2	2	2	2	2	
h	6	4	−32	−37	−41	−43	−1.6	h	10	2	1	1	1	1	
g	6	5	1	3	6	14	0.3	g	10	3	−5	−5	−5	−5	
h	6	5	−8	−6	−2	−2	0.5	h	10	3	2	3	3	2	
g	6	6	−111	−112	−111	−108	−0.1	g	10	4	−2	−1	−2	−2	
h	6	6	−7	1	11	17	0.0	h	10	4	6	4	4	5	
g	7	0	75	72	71	70	−1.0	g	10	5	4	6	5	5	
g	7	1	−57	−57	−56	−59	−0.8	h	10	5	−4	−4	−4	−4	
h	7	1	−61	−70	−77	−83	−0.4	g	10	6	4	4	4	3	
g	7	2	4	1	1	2	0.4	h	10	6	0	0	−1	−1	
h	7	2	−27	−27	−26	−28	0.4	g	10	7	0	1	1	1	
g	7	3	13	14	16	20	0.5	h	10	7	−2	−1	−1	−2	
h	7	3	−2	−4	−5	−5	0.2	g	10	8	?	0	0	2	
g	7	4	−26	−22	−14	−13	1.6	h	10	8	3	3	3	4	
h	7	4	6	8	10	16	1.4	g	10	9	2	3	3	3	
g	7	5	−6	−2	0	1	0.1	h	10	9	0	1	1	−1	
h	7	5	26	23	22	18	−0.5	g	10	10	0	−1	−1	0	
g	7	6	13	13	12	11	0.1	h	10	10	−6	−4	−5	−6	

gives rise to the driving electromotive force is known as the homogeneous dynamo problem. A number of detailed accounts of dynamo theory have been given (see e.g. Gubbins 1974, Busse 1978, Moffatt 1978, Roberts and Soward 1978, Krause and Rädler 1980) and no attempt will be made to review the subject in this book. Those aspects of the different models that have been proposed which have a bearing on the question of reversals of the field will be discussed later.

The dynamo problem involves the solution of a highly complicated system of coupled partial differential equations—electrodynamic, hydrodynamic and thermodynamic. Elsasser (1954) showed by a dimensional analysis that in geophysical and astrophysical problems the displacement current and all purely electrostatic effects are negligible, as are all relativistic effects of order higher than U/c where U is the fluid velocity. Thus the electromagnetic field equations are the usual Maxwell equations

$$\nabla \times \boldsymbol{E} = - \partial \boldsymbol{B}/\partial t \tag{1.19}$$

$$\nabla \times \boldsymbol{B} = \mu_0 \boldsymbol{j} \tag{1.20}$$

$$\nabla \cdot \boldsymbol{B} = 0 \tag{1.21}$$

where $\boldsymbol{B}$ and $\boldsymbol{E}$ are the magnetic and electric fields respectively and $\boldsymbol{j}$ the electric current density. The electromotive forces which give rise to $\boldsymbol{j}$ are due both to electric charges and to motional induction so that the total current $\boldsymbol{j}$ is given by

$$\boldsymbol{j} = \sigma(\boldsymbol{E} + \boldsymbol{U} \times \boldsymbol{B}). \tag{1.22}$$

Assuming the electrical conductivity σ to be constant, taking the curl of equation (1.20), and using equations (1.22) and (1.19), $\boldsymbol{E}$ can be eliminated, leading to the equation

$$\nabla \times (\nabla \times \boldsymbol{B}) = \mu_0 \sigma \left(-\frac{\partial \boldsymbol{B}}{\partial t} + \nabla \times (\boldsymbol{U} \times \boldsymbol{B}) \right). \tag{1.23}$$

Since $\nabla \times (\nabla \times \boldsymbol{B}) = \nabla(\nabla \cdot \boldsymbol{B}) - \nabla^2 \boldsymbol{B} = -\nabla^2 \boldsymbol{B}$, on using equation (1.21), we finally obtain

$$\partial \boldsymbol{B}/\partial t = \nabla \times (\boldsymbol{U} \times \boldsymbol{B}) + v_m \nabla^2 \boldsymbol{B} \tag{1.24}$$

where

$$v_m = 1/\mu_0 \sigma \tag{1.25}$$

is the magnetic diffusivity. Equations (1.21) and (1.24) give the relationships between $\boldsymbol{B}$ and $\boldsymbol{U}$ which have to be satisfied from electromagnetic considerations. The term $\nabla \times (\boldsymbol{U} \times \boldsymbol{B})$ in equation (1.24) is the source term, and represents the physical process by which magnetic induction is 'created' through the flow of fluid across lines of force. The term $v_m \nabla^2 \boldsymbol{B}$ represents the tendency for the field to decay through ohmic dissipation by the electric currents supporting the field. The balance between these two terms, at a particular point, determines how the magnetic field changes with time at that point.

If the material is at rest equation (1.24) reduces to

$$\partial \boldsymbol{B}/\partial t = v_m \nabla^2 \boldsymbol{B}. \tag{1.26}$$

This has the form of a diffusion equation, and indicates that the field leaks through the material from point to point. Dimensional arguments indicate a decay time of the order L^2/v_m where L is a length representative of the dimensions of the region in which current flows. For conductors in the laboratory this decay time is very small—even for a copper sphere of radius 1 m it is less than 10 s. For cosmic conductors on the other hand, because of their enormous size, it can be very large. As an alternative limiting case, suppose that the material is in motion but has negligible electrical resistance. Equation (1.24) then becomes

$$\partial \boldsymbol{B}/\partial t = \nabla \times (\boldsymbol{U} \times \boldsymbol{B}). \tag{1.27}$$

This equation is identical to that satisfied by the vorticity in the hydrodynamic theory of the flow of a non-viscous fluid where it is shown that vortex lines move with the fluid. Thus equation (1.27) implies that the field changes are the same as if the magnetic lines of force were 'frozen' into the material.

When neither term on the right-hand side of equation (1.24) is negligible, both the above effects are observed, i.e. the lines of force tend to be carried about with the moving fluid and at the same time leak through it.

If L, T, V represent the order of magnitude of a length, time and velocity respectively, transport dominates leak if $LV \gg v_m$. The condition for the onset of turbulence in a fluid is that the non-dimensional Reynolds number $R_e = LV/v$ be numerically large. By analogy, a magnetic Reynolds number R_m may be defined as

$$R_m = LV/v_m. \tag{1.28}$$

Thus the condition for transport to dominate leak is that $R_m \gg 1$. This condition is only rarely satisfied in the laboratory—in cosmic masses, however, it is easily satisfied because of the enormous size of L. Thus under laboratory conditions, lines of force slip readily through the material—in cosmic masses, on the other hand, the leak is very slow and the lines of force can be regarded as very nearly frozen into the material.

The generation of magnetic fields in astrophysical bodies relies on fluid motion having a large R_m—in the case of a dynamo operating in the Earth's core, it has been estimated that $R_m \gtrsim 10$. This is necessary so that the magnetic field can be distorted enough by the fluid motion to reinforce the large-scale field. Additionally, however, it is necessary that magnetic field lines diffuse sufficiently through the fluid and dissipate rapidly enough to keep the field topology simple. Otherwise the main effect of the fluid motion would simply be to tangle the field lines without producing efficient regeneration.

To the electromagnetic equations must be added the hydrodynamical equation of fluid motion in the Earth's core (the Navier–Stokes equation) together with the equation of continuity, which, for an incompressible fluid (the speed of flow is much less than the speed of sound in the Earth's core) reduces to

$$\nabla \cdot \boldsymbol{U} = 0. \tag{1.29}$$

The Navier–Stokes equation is

$$\rho \left(\frac{\partial \boldsymbol{U}}{\partial t} + (\boldsymbol{U} \cdot \nabla) \boldsymbol{U} + 2\boldsymbol{\Omega} \times \boldsymbol{U} - \nu \nabla^2 \boldsymbol{U} \right) - \frac{1}{\mu_0} (\nabla \times \boldsymbol{B}) \times \boldsymbol{B} = -\nabla p + \rho \nabla W \tag{1.30}$$

where $\boldsymbol{U}$ is the velocity relative to a system rotating with angular velocity $\boldsymbol{\Omega}$, p the pressure, W the gravitational potential (in which is absorbed the centrifugal force) and ρ and ν the density and kinematic viscosity, respectively. Equations (1.24) and (1.30) contain only the vectors $\boldsymbol{U}$ and $\boldsymbol{B}$ and are the basic equations of field motion.

There is no *a priori* reason why the Earth's magnetic field should have a particular polarity and there is no fundamental reason why its polarity should not change. It is easy to see that dynamos can produce a field in either direction. The induction equation (1.24) is linear and homogeneous in the field and the Navier–Stokes equation (1.30) inhomogeneous and quadratic. Thus, if a given velocity field will support either a steady or a varying magnetic field, then it will also support the reversed field and the same forces will drive it. This, however, merely shows that the reversed field satisfies the equations—it does not prove that reversal will take place.

Because of the complexity of the equations describing the hydromagnetic conditions in the Earth's core, most effort has been directed to seeking solutions of Maxwell's equations for a given velocity distribution. This approach, known as the kinematic dynamo problem, is linear and has been the subject of much investigation. Although it is now known that kinematic dynamos exist, solutions in which Maxwell's equations are solved for specified velocities are of limited geophysical interest since there is no guarantee that there exist forces in the Earth's core that can sustain them. Without a satisfactory theory to account for the driving force, the problem is not realistic and has only been 'pushed one stage further back'. In a dynamical theory the velocities would be calculated from assumed forces—very little work has been done on this more difficult non-linear problem. A number of possible driving mechanisms for the Earth's dynamo have been proposed (for a comprehensive review, see Gubbins and Masters (1979))—a general discussion of this problem is beyond the scope of this book.

The power source must replace the energy lost by electric currents and be efficient enough to do so without generating too much heat (the observed surface heat flux is about 4×10^{13} W, most of which comes from radioactive decay in the crust, with at most about 25 % or 10^{13} W coming from the core). The power source must also be long lasting since palaeomagnetic measurements have shown that the Earth's magnetic field has existed for at least 3500 Myr. Gubbins and Masters (1979) discuss in detail five possible power sources—radioactive heating, primordial heat and freezing of the inner core, gravitational energy release, chemical effects such as the heat of reaction between components, and precessional and tidal forcing. They favour stirring of the core by differentiation associated with the growth of the inner core. Shock wave data indicate that the outer core is predominantly iron with some 8–15 % light alloying elements and the inner core pure iron, with perhaps a little nickel. An advantage of this mechanism is its greater efficiency (see e.g. Gubbins 1977, Loper 1978, Gubbins *et al* 1979). All the gravitational energy released by rearrangement of core material is available to generate magnetic field. This process stirs the core directly, the magnetic field being the medium through which mechanical work is converted into heat. It must be stressed that the concentration of light material in the core is a different *component* in the mixture. Busse (1972) and Malkus (1973) have considered slurries in which the concentration is that of the solid phase. The essential difference is that the concentration of a phase can change rapidly because of pressure changes, since it only requires melting or freezing rather than actual molecular diffusion of one component.

Schloessin and Jacobs (1980) have suggested a modification to the above model which might also explain reversals of the Earth's magnetic field. In their model the light material rises all the way to the MCB and does not recombine at some higher level in the core. Thus the inner core and lower mantle have been growing from some time soon after the Earth formed at the expense of an initially more extended and probably entirely liquid core. On this assumption motions in the outer core are caused and sustained by currents which offset density inhomogeneities due to both changes in concentration of certain constituent components and to changes in temperature generated at the advancing mantle–core (MCB) and inner core (ICB) boundaries. In this model the energy available for fluid core motions and hence for maintenance of the magnetic field is directly related to the time rate of change of the growth of the solid components at the ICB and MCB. Magnetic polarity reversals might be explained as due to epochs during which the solid growth rate which dominates the fluid motion shifts from the ICB to the MCB and vice versa.

1.4 Mean field electrodynamics

Most dynamo models use large-scale, highly ordered fluid motions, i.e. motions in which the characteristic length of the velocity field is not much less than the radius of the Earth. In the early 1950s several attempts were made to produce models in which turbulent (i.e. random and small-scale) velocities might act as dynamos. The modern theory, which has been called mean field electrodynamics, has been developed independently by Moffatt (1970) in Britain and by Krause, Rädler and Steenbeck in Germany. An account of the German work has been given in a recent book by Krause and Rädler (1980).

In mean field dynamo models the velocity U and magnetic field B are each represented as the sum of a statistical average and a fluctuating part. We thus write

$$U = U_0 + u \qquad \langle u \rangle = 0 \qquad (1.31)$$

$$B = B_0 + b \qquad \langle b \rangle = 0. \qquad (1.32)$$

The average fields U_0 and B_0 are assumed to vary on a length scale L, while the fluctuating fields u and b (with zero statistical average) are assumed to vary on a length scale l ($l \ll L$). This separates the velocity and magnetic fields into mean, slowly varying and fluctuating parts. The induction equation may then be divided into its mean and fluctuating parts

$$\partial B_0/\partial t = \nabla \times (U_0 \times B_0) + \nabla \times \varepsilon + v_m \nabla^2 B_0 \qquad (1.33)$$

$$\partial b/\partial t = \nabla \times (U_0 \times b) + \nabla \times (u \times B_0) + \nabla \times G + v_m \nabla^2 b \qquad (1.34)$$

where

$$\varepsilon = \langle u \times b \rangle \qquad \text{and} \qquad G = u \times b - \langle u \times b \rangle. \qquad (1.35)$$

ε can be regarded as an extra mean electric force arising from the interaction of the turbulent motion and magnetic field. If the velocity field is isotropic, it can be shown that

$$\varepsilon = \alpha B_0 - \beta \nabla \times B_0 \qquad (1.36)$$

where α and β depend on the local structure of the velocity field.

The induction equation (1.33) satisfied by the mean field is

$$\partial B_0/\partial t = \nabla \times (\alpha B_0 + U_0 \times B_0) + (v_m + \beta) \nabla^2 B_0. \qquad (1.37)$$

The term αB_0 represents an electric field parallel to B_0. The quantity β is an eddy diffusivity similar in its effects to the ohmic diffusivity v_m. It operates by mixing magnetic fields transported from neighbouring regions—its effect is to replace v_m by a total diffusivity $v_T = v_m + \beta$. Parker (1955) first drew attention to the possibility that $\varepsilon = \alpha B_0$, and Steenbeck and Krause (1966) christened this the α effect. A key concept in the theory is the helicity defined as $u \cdot (\nabla \times u)$. Parker (1955, 1970) showed that convective fluid motions having non-zero helicity could distort lines of magnetic force in such a way as to produce a regeneration mechanism. A non-vanishing helicity indicates that the vorticity $\nabla \times u$ tends to turn predominantly either in a clockwise or anticlockwise sense about the direction of the velocity (for further details see Moffatt 1978, Krause and Rädler 1980). The helicity of fluid motions is a consequence of the action of the Coriolis force on convection in rotating bodies. This is the reason for the importance of rotation in the dynamo mechanism. It must be remembered, however, that the Coriolis force can only change the direction of flow—it cannot change the speed since it acts at right angles to the direction of flow, and so cannot drive it against the retarding effects of other forces.

There are two possible types of dynamo using the α effect—α^2 dynamos and $\alpha\omega$ dynamos. In an α^2 dynamo the α effect generates poloidal field from toroidal field and generates toroidal field from poloidal field. The toroidal field has lines of force that lie on spherical surfaces and has no component external to the core. The poloidal field has a radial component in general, and joins continuously with the external, observed field. The α effect can also be used in conjunction with a large-scale shear flow (the ω effect) to produce an $\alpha\omega$ dynamo. Parker (1955, 1971, 1979) is primarily responsible for the development of dynamo models of this type, in which an α effect from cyclonic turbulence generates poloidal field from toroidal field, and differential rotation creates toroidal field from poloidal field thereby completing the cycle. A good description of analytic and numerical work on $\alpha\omega$ dynamos has been given by Moffatt (1978). Gubbins (personal communication) has suggested that reversals of the magnetic field could arise in an $\alpha\omega$ dynamo through a change in direction of the differential rotation. This would destroy the poloidal field which would later grow in the opposite sense.

References

Bartels J 1936 The eccentric dipole approximating the Earth's magnetic field *Terr. Mag.* **41** 225

Bullard E C 1968 Reversals of the Earth's magnetic field *Phil. Trans. R. Soc.* A **263** 481

Bullard E C, Freedman C, Gellman H and Nixon J 1950 The westward drift of the Earth's magnetic field *Phil. Trans. R. Soc.* A **243** 67

Busse F H 1972 Comments on paper by G Higgins and G C Kennedy. 'The adiabatic gradient and the melting point gradient in the core of the Earth' *J. Geophys. Res.* **77** 1589

——1978 Magnetohydrodynamics of the Earth's dynamo *Ann. Rev. Fluid Mech.* **10** 435

Dagley P, Wilson R L, Ade-Hall J M, Walker G P L, Haggerty S E, Sigurgeirsson T, Watkins N D, Smith P J, Edwards J and Grasty R L 1967 Geomagnetic polarity zones for Icelandic lavas *Nature* **216** 25

Elsasser W M 1954 Dimensional values in magnetohydrodynamics *Phys. Rev.* **95** 1

Gubbins D 1974 Theories of the geomagnetic and solar dynamos *Rev. Geophys. Space Phys.* **12** 137

——1977 Energetics of the Earth's core *J. Geophys.* **43** 453

Gubbins D and Masters T G 1979 Driving mechanisms for the Earth's dynamo *Adv. Geophys.* **21** 1

Gubbins D, Masters T G and Jacobs J A 1979 Thermal evolution of the Earth's core *Geophys. J.* **59** 57

Hide R 1966 Free hydromagnetic oscillations of the Earth's core and the theory of geomagnetic secular variation *Phil. Trans. R. Soc.* A **259** 615

Hide R and Malin S R C 1970 Novel correlations between global features of the Earth's gravitational and magnetic fields *Nature* **225** 605

Krause F and Rädler K H 1980 *Mean Field Magnetohydrodynamics and Dynamo Theory* (Oxford: Pergamon)

Loper D E 1978 The gravitationally powered dynamo *Geophys. J.* **54** 389

McDonald K L and Gunst R H 1967 An analysis of the Earth's magnetic field from 1835 to 1965 *IERT Tech. Rep.* 46-IESI (Boulder, Colorado)

Malkus W V R 1973 Convection at the melting point, a thermal history of the Earth's core *Geophys. Fluid Dyn.* **4** 267

Moffatt H K 1970 Turbulent dynamo action at low magnetic Reynolds number *J. Fluid Mech.* **41** 435

——1978 *Magnetic Field Generation in Electrically Conducting Fluids* (Cambridge: Cambridge University Press)

Moffatt H K and Dillon R F 1976 The correlation between gravitational and geomagnetic fields caused by interaction of the core fluid motion with a bumpy core–mantle interface *Phys. Earth Planet. Int.* **13** 67

Nelson J H, Hurwitz L and Knapp D G 1962 Magnetism of the Earth *US Department of Commerce Coast Geod. Surv.* Publ. 40-1

Opdyke N D, Glass B, Hays J D and Foster J 1966 Palaeomagnetic study of Antarctic deepsea cores *Science* **154** 349

Parker E N 1955 Hydromagnetic dynamo models *Astrophys. J.* **122** 293

——1970 The generation of magnetic fields in astrophysical bodies. I. The dynamo equations *Astrophys. J.* **162** 665

——1971 The generation of magnetic fields in astrophysical bodies. IV. The solar and terrestrial dynamos *Astrophys. J.* **164** 491

——1979 *Cosmical magnetic fields, their origin and their activity* (Oxford: Clarendon)

Peddie N W 1982 International Geomagnetic Reference Field: the third generation *J. Geomag. Geoelec.* **34** 309

Roberts P H and Soward A M 1978 *Rotating fluids in geophysics* (New York: Academic)

Schloessin H H and Jacobs J A 1980 Dynamics of a fluid core with inward growing boundaries *Can. J. Earth Sci.* **17** 72

Skiles D D 1970 A method of inferring the directions of drift of the geomagnetic field from palaeomagnetic data *J. Geomag. Geoelec.* **22** 441

——1972a The laws of reflection and refraction of incompressible magnetohydrodynamic waves in a fluid–solid interface *Phys. Earth Planet. Int.* **5** 90

——1972b On the transmission of the energy in an incompressible magnetohydrodynamic wave into a conducting solid *Phys. Earth Planet. Int.* **5** 99

Steenbeck M and Krause F 1966 The generation of stellar and planetary magnetic fields by turbulent dynamo action *Z. Naturf.* a **21** 1285

Vestine E H, Laporte L, Cooper C, Lange I and Hendrix W 1947 Description of the Earth's main magnetic field and its secular change, 1905–1945 *Carnegie Inst. Wash. Publ.* No 578

Yukutake T 1962 The westward drift of the magnetic field of the Earth *Bull. Earthquake Res. Inst.* **41** 1

Yukutake T and Tachinaka H 1969 Separation of the Earth's magnetic field into drifting and standing parts *Bull. Earthquake Res. Inst.* **47** 65

2 The magnetisation of rocks

2.1 Introduction

Although most rock-forming minerals are non-magnetic, all rocks show some magnetic properties due to the presence of various iron oxide and sulphide minerals making up only a few per cent of the rock. These minerals occur as small grains dispersed through the magnetically inert matrix provided by the more common silicate minerals that make up most rocks. When a rock forms it usually acquires a magnetisation parallel to the ambient magnetic field referred to as a *primary magnetisation*. This can give information about the direction and intensity of the magnetic field in which the rock formed. However, subsequent to formation, the primary magnetisation may decay either partly or wholly and further components may be added by a number of processes. These subsequent magnetisations are called *secondary magnetisation*. The magnetisation first measured in the laboratory is called the *natural remanent magnetisation* (NRM). A major problem in palaeomagnetic investigations is to recognise and eliminate secondary components.

The mineralogy of rock magnetism is complicated by a multiplicity of phases and solid solutions of iron oxide, particularly with titanium dioxide. Most magnetic minerals are within the ternary systems FeO–Fe_2O_3–TiO_2. There are essentially two types of mineral: the strongly magnetic cubic oxides magnetite (Fe_3O_4), maghemite (γFe_2O_3) and the solid solutions of magnetite with ulvospinel (Fe_2TiO_4) known as titanomagnetite, and the more weakly magnetic, rhombohedral minerals based on hematite (αFe_2O_3) and its solid solutions with ilmenite ($FeTiO_3$). The physical properties of rock magnetism have been discussed in detail by Fuller (1970), Stacey and Banerjee (1974) and Creer *et al* (1975).

There are several different types of magnetism. In a typical substance there is no overall magnetism arising from the orbital motion of the electrons round the central nucleus. However, if it is placed in a magnetic field, a force is exerted on each of the orbital electrons tending to modify its orbit slightly. Since the resulting effect tends to oppose the applied field, the magnetisation acquired in this way is negative (i.e. the susceptibility is negative) and is called *diamagnetism*. All substances possess diamagnetism which is independent of temperature, although it is often obscured by other superimposed effects. Many common minerals such as quartz and feldspar are predominantly diamagnetic. The electrons also have a spin motion about their axes in addition to their orbital motion around the central nucleus. They are usually randomly oriented, but in the presence of a magnetic field they tend to line up in the direction of the field, giving an increase in the magnetisation. This effect is called *paramagnetism*. In natural minerals, only a few important ions show significant paramagnetic properties, the commonest being Mn^{2+}, Fe^{3+} and Fe^{2+}. Since an applied magnetic field tends to orient the spins while thermal fluctuations tend to randomise them, paramagnetic effects are strongly dependent on temperature.

Some materials exhibit a permanent (spontaneous) magnetisation even in the absence of an external field. This phenomenon of *ferromagnetism* is due to the 'exchange' interaction between atoms. Some substances (e.g. iron, cobalt, nickel) contain unpaired electrons which are magnetically coupled between neighbouring atoms. This interaction results in a strong spontaneous magnetisation (i.e. without the application of an external field) and in the property of being able to retain the alignment imparted by an applied field after it has been removed. These properties are several orders of magnitude greater than those of diamagnetic atoms in the same substance. Below some critical temperature (the Curie temperature) the interaction dominates, but above this temperature, thermal disordering takes over and the behaviour is that of a simple paramagnetic material.

Some substances are characterised by subdivision into two sublattices, the atomic moments of which are each aligned, but antiparallel to one another. The ferromagnetic effects cancel one another out when the moments of the two sublattices are equal and there is no net magnetic moment. This phenomenon is known as *antiferromagnetism*. Such substances do not have a Curie temperature, because there is no net ferromagnetism. In this case the ordering of the atomic moments is destroyed at a critical temperature called the Néel temperature, above which the substances exhibit paramagnetism. If the atomic moments of the two sublattices are unequal, there is a net spontaneous magnetisation—a weak ferromagnetism known as *ferrimagnetism*. Most naturally occurring magnetic minerals are either antiferromagnetic or ferrimagnetic. Ferrimagnetic minerals include magnetite, maghemite, and some members of the ilmenite–hematite solid solution series; antiferromagnetic minerals include hematite, ilmenite, and ulvospinel. These last two minerals, however, have Curie or Néel temperatures well below room temperature. Some minerals possess a feeble spontaneous magnetisation which is superposed on an antiferromagnetic structure and which disappears along with the antiferromagnetism at the Néel temperature. This is called imperfect antiferromagnetism and may be due to imperfect antiparallel alignment (canting),

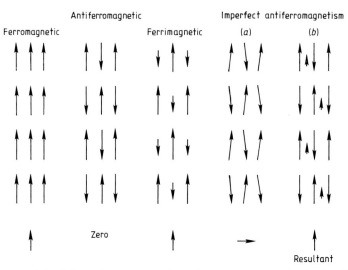

Figure 2.1   Schematic representation of spontaneous magnetisation in crystals. The arrows represent the elementary moments, the resultants being given at the bottom (after Irving 1964).

lattice imperfections or to a small parasitic component (see figure 2.1(*a*), (*b*)). The best known canted antiferromagnetic is hematite in which equal sublattice magnetisations are not quite antiparallel so that there is a small net magnetisation normal to the average spin axis.

A certain amount of energy (magnetostatic energy) is required for a body to exhibit remanence. While the exchange energy tends to line up all the spins, the magnetostatic energy attempts to prevent this in order to minimise the total energy. The result is a balance of energy in which small zones, a few microns in size, are uniformly magnetised while adjacent zones may have their magnetisation in some other direction. These zones are called *magnetic domains*, domain walls separating adjacent regions in which the spontaneous magnetisation is in different directions. A sample of ferromagnetic material may thus have only a weak overall spontaneous magnetisation, although in individual domains it may be quite large.

Within any crystal there are certain 'easy' directions of minimum energy along which the magnetic dipoles prefer to be aligned. However, in domain walls separating adjacent regions of uniform magnetisation, some spins cannot be aligned in the preferred magnetocrystalline direction; these particular spins require extra energy to overcome the magnetocrystalline effect. In the absence of any external influence, domains form in such a manner as to reduce the total magnetostatic energy and wall energy to a minimum. When a body is placed in a magnetic field, the domain walls can move fairly easily allowing more of the grain to become magnetised in the direction of the applied field. Such magnetisation is reversible when the applied field is small, and the domain walls move back again when the field is removed. If the field is increased, the domain walls are forced over small imperfections and impurities in the grain and cannot return to their original position when the field is removed. The process is then no longer reversible and a definite permanent magnetisation is left in the body. If higher fields still are applied, all the spins line up in the direction of the applied field, overcoming both the magnetostatic and magnetocrystalline energies. At this point the body has spontaneous or saturation magnetisation. For small grains no domain walls can occur (single domain grains). At some critical size, the grain will subdivide into two or more domains and form multidomain grains.

The magnetic behaviour of single domain and multidomain grains is quite different and it is important to know which configurations are relevant to the magnetic minerals in rocks. Hematite commonly behaves as single domained while magnetite grains are generally multidomained. Some small multi-domained grains (diameter $\leqslant 15$ μm) have some single domain properties, including high stability of remanence which is of particular importance in palaeomagnetic studies. Such grains are referred to as pseudo-single domain. The basic reason for such behaviour is due to the interaction of a domain wall with crystal defects so that none of its possible stable positions coincides with the position which gives zero total magnetic moment for the grain. The essential features of the behaviour of single domain grains in rocks over geologic time can be adequately described in terms of Néel's (1949, 1955) simple theory.

The mechanism by which NRM is acquired depends upon the mode of formation and subsequent history of the rocks as well as the characteristics of the magnetic minerals they contain. Rocks may become magnetised by a number of different natural processes.

Thermoremanent magnetisation (TRM)

Igneous rocks cooling from above their Curie temperature in a magnetic field acquire a remanent magnetisation, called thermoremanent magnetisation (TRM). This magnetisation is parallel to the applied field and for low field strengths is directly proportional to it. If a rock is cooled through various temperature intervals in the presence of a magnetic field, the TRM acquired in each interval is found to be independent of that acquired in each of the other intervals, a result known as the principle of partial thermoremanent magnetism (PTRM) superposition. The sum of all the PTRM values is equal to the total TRM acquired by cooling from above the Curie temperature to room temperature.

Isothermal remanent magnetisation (IRM)

Magnetic materials may acquire a remanent magnetisation without heating if exposed to a magnetic field. In many natural materials, the IRM acquired in the

Earth's magnetic field is less than the TRM, although this may not be true for magnetically soft materials.

Viscous remanent magnetisation (VRM)

Viscous remanent magnetism is essentially time-dependent IRM. If a magnetic material is exposed to a magnetic field it may slowly acquire a magnetisation in the direction of that field. The rate at which VRM is acquired (or decays) depends on the temperature and particle size. For any particle the relaxation time $\tau \propto \exp(v/T)$ where v is the volume of the grain and T the temperature. Very fine ferromagnetic particles (less than domain size) have very short relaxation times even at normal temperatures and are said to be *superparamagnetic*. A small increase in volume can change a particle from superparamagnetic to single domain; a small decrease in temperature leads to the same effect. A material which contains many small particles will have some which are superparamagnetic and some which are single domain. For each grain there is a critical blocking temperature T_B at which τ becomes small but which might also be below the Curie temperature. Similarly at any given temperature T, there is a critical blocking diameter d_B (corresponding to a sphere of volume v_B) at which τ becomes small.

Depositional remanent magnetisation (DRM)

Some sediments contain small magnetic particles which become aligned in the direction of the ambient magnetic field as they fall through the water. A post-depositional remanent magnetisation (PDRM) may be acquired if the carrier grains are free to rotate in the voids of the sediment matrix as they attempt to follow the secular variation of the geomagnetic field which occurred subsequent to deposition. The PDRM process has been studied experimentally in several laboratories during recent years and is now generally accepted as being the more important. In particular, Tucker (1980) has shown that the efficiency of the magnetisation process depends on the relative diameters of the carrier and bulk matrix grains, being most effective when the carrier grains are smaller than the matrix grains. The size of the voids progressively decreases as a consequence of compaction which occurs due to the deposition of more and more sediment as time goes by. Thus the ability of the carrier grains to rotate towards the geomagnetic field direction is progressively impeded and decreases sharply at a critical value of the porosity. PDRM is thought to be primarily responsible for the NRM of deep-sea sediments, with either DRM or PDRM for that of lake sediments.

Chemical remanent magnetisation (CRM)

Some sediments undergo post-depositional chemical changes in the iron minerals introducing a subsequent magnetisation. Such a process is referred to as chemical remanent magnetisation (CRM).

In the majority of igneous rocks the NRM consists mainly of TRM, even though CRM and VRM may often contribute significantly. The NRM is carried primarily by titanomagnetites—and sometimes by titanohematites. The magnetic mineralogy of deep-sea sediments is not well known; the NRM appears to be carried mainly by titanomagnetites though contributions from pyrrhotite (in certain anoxic environments) and ferromanganese oxides and oxyhydroxides also contribute. The magnetic mineralogy of terrestrial sediments is highly variable since the sediments originate in part from pre-existing igneous, sedimentary and metamorphic rocks and also because magnetic minerals sometimes form in sediments by authigenesis and diagenesis. The origin of the magnetisation of the so-called 'red beds' (principally red sandstones and shales) is discussed briefly later in this section.

The simplest magnetisation to interpret is that carried by a lava. The lavas cool through the range of blocking temperatures, in which the primary NRM is acquired as a TRM in a time which is short compared with the time constants of the secular variation. Hence each lava gives a spot reading of the local field direction. Determinations of the ancient field intensity are particularly reliable from such rocks. Unfortunately, the detailed chronology of lava flows is not usually sufficiently well known to time the sequence of events in a magnetic reversal.

Sedimentary records have the advantage that sedimentation is in general a more continuous process than is the extrusion of lavas. However, the type of sedimentary environment which gives the most continuous sedimentation is also one in which absolute sedimentation rates are low, so that resolution of the record tends to be poor. Another problem is that it is often difficult to know when a hiatus in sedimentation has occurred.

Channell (1978) has examined the magnetic properties in a bed of pelagic limestone of Cretaceous age from Sicily, which vary in colour from red to white. Samples taken from the red portion of the bed showed normal polarity, and those from the white, reversed polarity, the directions being almost antiparallel after partial demagnetisation. The remanence in the case of the white variety is due to detrital magnetite, and that in the red to hematite. The hematite magnetisation significantly postdates the detrital magnetisation. Fujiwara and Ohtake (1975) carried out a palaeomagnetic study of Late Cretaceous alkaline rocks from Hokkaido, Japan. Both normal and reversed polarities were found in one complex of picritic dolerites. They suggested that the existence of two different palaeomagnetic polarities in one rock complex might be explained on the assumption that the geomagnetic field had been normal at the time when the rock at the margins cooled through the Curie temperature of the ferromagnetic minerals in the rock, but had reversed its polarity when the rock inside the intrusive body became magnetised during cooling. This would imply that the time of the polarity transition was quite short since it occurred during the cooling of a relatively thin intrusive body.

In recent years there has been much controversy over the interpretation of palaeomagnetic data obtained from red beds (see e.g. Walker *et al* 1981). Many believe that red bed remanence faithfully mirrors the geomagnetic field present at (or soon after) the time of deposition. According to this view, remanence data from red beds can be used to study both large-scale as well as small-scale features of the ancient geomagnetic field. Elston and Purucker (1979) believe that the rapid acquisition of remanence is due to depositional remanence resulting from physical alignment of detrital hematite grains. On the other hand Helsley and Steiner (1974) suggest that it is due to chemical remanence, as a by-product of hematite authigenesis occurring within several thousand years after deposition. Purucker *et al* (1980) suggest that both processes may have occurred. Larson *et al* (1982), on the other hand, believe that none of the above explanations is viable and that red bed remanence primarily consists of multiple, commonly near-antiparallel components acquired through chemical alteration discontinuously over long time intervals. If this view is correct, the remanence in red beds rarely reflects the geomagnetic field at the time of deposition.

In 1975 a detailed study of the palaeomagnetic, rock magnetic, and petrographic properties of the Triassic Moenkopi Formation was carried out. The results of the petrographic study were published by Walker *et al* (1981), and the rock magnetic and palaeomagnetic study by Larson *et al* (1982). Larson *et al* found both normal and reversed Triassic components in many of the same samples at many localities and that there was a close correlation between the magnetic polarity and intensity of magnetic remanence and the lithologic characteristics in the Sinbad Valley section. Moreover the intermediate remanence directions throughout the formation were of weak intensity suggesting the presence of two or more nearly balanced antiparallel components. They concluded that the principal remanence carried by the formation is CRM acquired diagenetically over a geologically long time interval, and they presented a model to explain how it was acquired through natural processes acting intermittently at geologically reasonable rates.

Denham (1981) has examined the effect of field reversals on the length of time a rock can preserve a particular polarity. The longevity of a magnetic polarity message depends on the decay time constants of the constituent magnetic grains and Denham estimated the minimum time constants necessary for the preservation of an ancient polarity message against the alternating sequence of polarity reversals. He showed that the viscous decay of a magnetic polarity message in rocks is slowed by a factor of three to six due to reversals of the field compared with the effect in a field that is constantly polarised opposite to the original remanence. It should be noted, however, that increases in temperature have a far greater effect than reversals of polarity, e.g. for original time constants of 1–100 Myr, halving and quartering the relaxation time has the same effect as raising the ambient temperature from 273 K to nominal values of only 276.5 K and 280 K respectively. Raising the temperature to 300 K would lower the time constants by more than two orders of magnitude.

2.2 Self-reversal

Although evidence (to be discussed later) is overwhelmingly in favour of reversals of the Earth's field as the cause of reversed magnetisation, before such an explanation is accepted, it must be asked whether any physical or chemical processes exist whereby a material could acquire a magnetisation opposite in direction to that of the ambient field. Graham (1949) found some sedimentary rocks of Silurian age which were reversely magnetised. He was able to identify the precise geological horizon over a distance of several hundred kilometres by the presence of a rare fossil which existed only during a short geological period. He found that some parts of the horizon were normally magnetised and some reversely, and argued that this could not be accounted for by a reversal of the Earth's field which would affect all contemporaneous strata alike. (However, if the timescale of reversals in Silurian times was as short as has been found in some later rocks, Graham's fossil might well have survived at least one reversal.) Graham thus wrote to Professor Néel at Grenoble and asked him if he could think of any process by which a rock could become magnetised in a direction opposite to that of the ambient field. Néel (1951, 1955) suggested, on theoretical grounds, four possible mechanisms.

The first and third of Néel's mechanisms involve only reversible physical and/or chemical changes. In his first mechanism Néel imagined a crystalline substance with two sublattices A and B, with the magnetic moments of all the magnetic atoms in lattice B oppositely directed to those of lattice A. If the spontaneous magnetisation of the two sets of atoms J_A and J_B vary differently with temperature, Néel suggested that the resultant magnetisation of the whole, $J_A + J_B$, could reverse with change in temperature (figure 2.2). Gorter and Schulkes (1953) later synthesised a range of substances with the properties predicted by Néel, although no naturally occurring rock has been found which behaves in this manner. In ferrites, the two constituents A and B are the two interpenetrating cubic sublattices which make up the crystal structure. Because the exchange interaction between the cation sites in the two sublattices is negative, the B sublattice acquires a spontaneous magnetisation exactly antiparallel to that of the A sublattice when the ferrite cools through its Curie temperature. Given appropriate temperature coefficients for the two spontaneous magnetisations, reproducible self-reversal will occur on further cooling.

Néel's second mechanism is a modification of the first, in which $J_A > J_B$ at all temperatures, so that no reversal would take place. However Néel suggested that subsequent to the formation of such a substance, chemical or physical

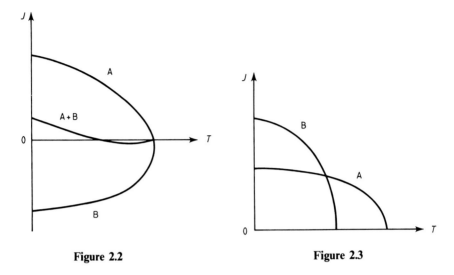

Figure 2.2   Figure 2.3

changes might occur which would lead to the demagnetisation of lattice A, leaving the reverse magnetisation of lattice B predominant. No evidence of such a possibility actually occurring in nature has yet been found.

In his third mechanism, Néel considered a substance containing a mixture of two different types of grains, A and B, one with a high Curie point T_A and a low intensity of magnetisation J_A, and the other with a low Curie point T_B and a high magnetisation J_B (figure 2.3). When such a substance cools from a high temperature, substance A, because of its higher Curie point, becomes magnetised first in the direction of the ambient field. When the temperature falls below T_B, substance B becomes magnetised but will be subject to the dual influence of the ambient field and of the field due to the grains of substance A. Néel suggested that under suitable geometrical conditions, the resultant direction of magnetisation of B could on the average be opposite to that of the ambient field. At room temperature the greater value of J_B causes the resultant magnetisation of the whole to be in the opposite direction to the ambient field.

Néel's fourth mechanism, like his second, involves the possibility of subsequent demagnetisation by physical or chemical changes. The reverse magnetisation might be possible later in time, even though initially the intensity of the B component was not large enough. Since discrete magnetised grains free to rotate must align themselves along and not against the field, only the second and fourth of Néel's mechanisms could apply to sedimentary rocks; igneous rocks could, in theory, become reversely magnetised by any of them.

In 1952 Nagata *et al* found a dacite pumice from Mt Haruna, Japan, which was shown in the laboratory to be self-reversing and to contain an intimate mixture of two magnetic minerals, an ilmenohematite and a titanomagnetite, indicating reversal by magnetostatic interaction (Nagata *et al* 1952, Uyeda

1955). However, subsequent separation of the minerals revealed that the self-reversal was an intrinsic property of one of them alone and that the presence of titanomagnetite was irrelevant (Uyeda 1958, Nagata and Uyeda 1959). Ishikawa and Syono (1963) showed that the reversal is connected with the ordering and disordering of iron and titanium ions in the lattice. Self-reversal occurs only when ordered and disordered phases are both present in metastable equilibrium, and seems to be the result of an antiparallel superexchange coupling between the phases. Specimens which are completely ordered or disordered did not show reversed TRM. The mechanism is essentially different from any of Néel's mechanisms. Moreover, the self-reversing properties were a function of cooling rate. Since the rate of cooling in laboratory experiments is many orders of magnitude greater than that at which rocks cool in nature, the results may have no geophysical significance. Westcott-Lewis and Parry (1971) later found that the self-reversing property of ilmenohematites disappeared for grains smaller than 4 µm, which acquire only normal TRM. They suggested that this is because chemical inhomo-geneities become important in small grains which do not contain sufficient of both the ordered and metastable phases. Carmichael (1959, 1961) has also produced a self-reversal in the laboratory in an ilmenite–hematite solid solution containing much less ilmenite. He suggested that the self-reversal may be related to the exchange of electrons between Fe^{2+} and Fe^{3+} ions in the oppositely directed magnetic sublattices of the solid solution.

Heller and Egloff (1974) have examined the NRM of a granite–aplite dyke of the Bergell Massif, Switzerland, which showed stripes of normal and reversed polarity. Thermal experiments indicated 'imperfect' self-reversal due to the presence of two interacting phases with different Curie points within the exsolved ilmenohematite. Heller later (1980) carried out thermomagnetic experiments on the Olby–Laschamp lavas from the Chaîne des Puys, France, and showed that some of the samples of the Olby flow undergo complete self-reversal during thermal demagnetisation—other samples showed partial self-reversal of the NRM at room temperature. He suggested that the thermomag-netic behaviour is due to magnetic phases of titanomagnetite with different Curie temperatures. Heller and Petersen (1982a, b) later carried out magnetic and optical examination of some of the samples and suggested that the self-reversal is caused by a negative magnetostatic coupling between titanomag-netite phases of widely varying degrees of oxidation. (The magnetisation of the Olby–Laschamp lava flows is discussed further in chapter 4.)

The duration of the formation of the magnetic phases is highly important. If the interacting phases were present prior to the acquisition of remanence, then in a cooling lava a high-Curie-point phase would be magnetised first and the magnetisation of this phase would control the interactions leading to a reversal during further cooling. If the secondary interacting phases were formed after the acquisition of remanence, by low-temperature physical or chemical processes, then a primary phase having a low Curie point might be responsible for the reversal of a later phase having a higher Curie temperature.

Heller and Petersen (1982b) showed that the titanomagnetites of the Olby and Laschamp flows were oxidised under different conditions and to a varying extent. They suggested that the observed self-reversal is caused by interaction of two magnetic phases: little oxidised or unoxidised primary titanomagnetite having T_C around 180 °C, and secondary titanomaghemite having T_C up to 400 °C.

Kennedy (1981) has also found self-reversal of TRM in a dacite—from Mt Natib in the Philippines which may be explained by Ishikawa and Syono's (1963) model. The Fe-rich metastable phase produced during ordering in the minerals of the ilmenite–hematite solid solution represents the initial stages of the exsolution process. Exsolution-unmixing of an Fe-enriched phase aids ordering in the remainder of the mineral by increasing its Ti content.

Lawson *et al* (1981) have examined synthetic ilmenite-hematite samples by transmission electron microscopy and found well-defined antiphase domains (APD) and antiphase domain boundaries (APB) (APD are chemical domains and, as such, are distinct from magnetic domains). Samples synthesised or annealed at 900 °C, which is below the order–disorder transition for $Ilm_{70}Hem_{30}$, have much larger domains or no domains at all and consequently have a much smaller volume of APB than samples synthesised or annealed at 1300 °C. Only the high-temperature samples acquire reverse TRM when cooled in an applied magnetic field. Lawson *et al* hypothesised that some critical volume of APB is necessary for the acquisition of reverse TRM. In samples containing less than the critical volume, the APB are unable to couple magnetically with the main body of the domains effectively, and a normal TRM results.

Although examples of Néel's first mechanism (figure 2.2) have been found in ternary ferrites (Gorter and Schulkes 1953), the compositions are unlike those found in normal rocks. Two examples of this type of ferrimagnetism have however been reported. Schult (1968) found self-reversal of NRM in basalt specimens from Germany when cooled below 200 K; at the temperature of inversion the saturation magnetisation showed a minimum. No specimens were found to give self-reversals above about 200 K. The only example of this type of behaviour reproducible in the laboratory with an inversion temperature above 300 K was reported by Kropaček (1968) for a tin-substituted hematite embedded in a cassiterite rock.

Apart from the Haruna dacite a few other examples have been found of self-reversal of the NRM in volcanic rocks—in certain oceanic basalts (Ozima and Ozima 1967, Sasajima and Nishida 1974) and in a basalt from Germany (Schult 1976). A number of cases of partial reversal have also been discovered. A dacite pitchstone from Mount Asio, Japan (Nagata *et al* 1953) and an iron sand from Sokoto, Japan (Uyeda 1958) acquire a reversed TRM between 200°C and 300°C, but because it is small compared with that acquired parallel to the field in other temperature intervals, the total TRM is directed along the field. Everitt (1962) first demonstrated the self-reversal behaviour of pyrrhotite in the laboratory. Later Bhimasankaram (1964) showed that reversal in natural and synthetic pyrrhotites takes place between two components with Curie temperatures 560 °C and 310 °C coupled antiparallel to each other. Robertson (1963) has also discovered self-reversal in pyrrhotite in a monzonite from Australia.

Other partial self-reversals have been detected in pyrrhotite-bearing rocks and inferred in a few other cases. However, pyrrhotite is not common in rocks and is absent from most reversely magnetised rocks. Partial self-reversal leading only to a reduction of the NRM intensity at room temperature, but not to a completely antiparallel alignment of the NRM direction with respect to the ambient magnetic field, has been observed in an historical lava flow from Mount Etna (Heller *et al* 1979). It has also been shown that the TRM produced in continental basalts under moderate or high temperature oxidation conditions in the laboratory can acquire self-reversal or at least partial self-reversal characteristics at room temperature (Havard and Lewis 1965, Creer and Petersen 1969, Creer *et al* 1970, Petersen and Bleil 1973, Ryall and Ade-Hall 1975, Petherbridge 1977, Tucker and O'Reilly 1980).

Hoffman (1982) has analysed some Oligocene basalt samples, cored from Yarraman Creek in the Liverpool Range, New South Wales, Australia, and found that they contained two remanence components with opposing directions. The self-reversed component was clearly identified in several samples subjected to detailed thermal demagnetisation and was most evident in samples showing the highest degree of low-temperature oxidation. Several flows under investigation were extruded during what appears to have been an excursion of the geomagnetic field, each possessing a magnetisation direction far from the full polarity state. The possibility that the self-reversed moments are simply secondary components, acquired in an opposing field direction is, therefore, ruled out. Optical examinations and thermomagnetic curves indicated that the samples contain from nearly unoxidised to moderately low-temperature oxidised titanomagnetite. The self-reversed component is most clearly removed during thermal demagnetisation between 230 °C and 290 °C above the observed Curie point of the titanomagnetite, and is believed to be associated with titanomaghemite.

2.3 Field reversal or self-reversal?

To prove that a reversed rock sample has become magnetised by a reversal of the Earth's field, it is necessary to show that it cannot have been reversed by any physico-chemical process. This is almost impossible to do since physical changes may have occurred since the initial magnetisation or may occur during laboratory tests. More positive results can only come from the correlation of data from rocks of various types at different sites and by statistical analyses of the relation between the polarity and other chemical and physical properties of the rock sample.

There have been many cases where reversely magnetised lava flows cross sedimentary layers. Where the sediments have been baked by the heat of the cooling lava flow, they are also found to be strongly magnetised in the same reverse direction as the flow. In fact, in about 95 % of all cases the direction of magnetisation of the baked sediment is the same as that of the dike or lava which heated it, whether normal or reversed. It seems improbable that the adjacent rocks as well as the lavas themselves should possess a self-reversal property, and such results seem difficult to interpret in any other way than by a reversal of the Earth's field.

The same pattern of reversals observed in igneous rocks has also been found in deep-sea sediments (see e.g. figure 3.2). No two substances could be more different or have more different histories than the lavas of California and the pelagic sediments of the Pacific. The lavas were poured out, hot and molten, by volcanoes and magnetised by cooling in the Earth's field; the ocean sediments on the other hand accumulated grain by grain by slow sedimentation and by chemical deposition in the cold depths of the ocean. If these two materials show the same pattern of reversals then it must be the result of an external influence working on both and not due to a recurrent synchronous change in the two materials. The evidence seems compelling that reversals of the Earth's field are the cause of the reversals of magnetisation.

Reversals have been found during the Precambrian and have been observed in all subsequent periods. There is no evidence that periods of either polarity are systematically of longer or shorter duration. However during the Kiaman—a period of about 60 Myr within the upper Carboniferous and Permian (about 235–290 Myr ago)—the polarity of the Earth's field appears to have been almost always reversed: until quite recently no normal intervals at all were known within this period. McElhinny (1969) and Burek (1970) have now both reported a normal event at about 280 Myr and Creer *et al* (1971) another at about 263 Myr. If the field reversal hypothesis is incorrect, it follows that mineral assemblages necessary for self-reversal were abundant in Carboniferous and Triassic rocks (both these periods have many reversals), but were all but missing in all Permian rocks. Such a conclusion is very difficult to believe; it is far more plausible to assume that the field itself alternated very rarely during the Permian.

If the dipole field of the Earth has reversed, it is most probably a result of physical processes occurring in the core of the Earth, and should thus be quite uncorrelated with physical processes associated with the crust and upper mantle or the atmosphere, such as orogenic and volcanic activity or climatic changes. It should be pointed out however that the lack of correlation between a rock's polarity and other properties does not necessarily rule out self-reversal because the self-reversal mechanism may be intrinsically undetectable.

A number of workers have reported chemical differences between normally and reversely magnetiscd lava sequences from various parts of the world. The ferromagnetic minerals in reversely magnetised lavas appear to be more highly oxidised than those which are normally magnetised. This was first noted by Balsey and Buddington (1954, 1958) in the Precambrian rocks of the Adirondacks; it has since been confirmed by several workers in rocks from Mull, Siberia, Iceland, Japan and elsewhere (Wilson and Watkins 1967). The minerals concerned are mixed oxides of iron and titanium. The chemical evidence is that iron is in the high valency (Fe^{3+}) state more frequently in reversed rocks; the petrological evidence is that the iron–titanium oxide crystals are visibly in different, more highly oxidised, crystal form, while some of the silicate minerals are noticeably more reddened.

Larson and Strangway (1966) have argued that the correlation between magnetic polarity and petrology is probably fortuitous, and cite extensive investigations of basaltic samples from Oregon, New Mexico, California and Japan in which there was no such correlation. Wilson (1966a), however, continued to find distinct petrological differences between normal and reversed Tertiary basalts from Japan and Iceland and Carboniferous lavas from Scotland.

The Columbia Plateau basalts, which are of Miocene age, have been studied in more detail than any other sequence (Wilson and Watkins 1967). Here there are five series each of up to twenty lavas with reversed magnetisation which are predominantly but not exclusively highly oxidised alternating with five sequences with normal magnetisation and a predominantly low degree of oxidisation. The correlation of polarity with state of oxidisation (figure 2.4) is remarkable, particularly since the Columbia Plateau basalts show independent evidence that the reversed samples are due to field reversal (baked–baking pairs of the same polarity). A polarity–oxidation correlation seems undeniable, although it is not a one-to-one but a statistical correlation. There are highly oxidised rocks that are normally magnetised and reversely magnetised ones that are lightly oxidised. Not all lava sequences show as clear a correlation as is shown in figure 2.4; in fact, Watkins and Haggerty (1968) have shown that the state of oxidation often varies greatly in a single flow, being higher in the centre than near the upper or lower margins. The variations in oxidation appear to be an original feature of the flows and to have been produced during the original solidification and cooling, probably by oxygen derived from the decomposition of water. Larson and Strangway (1968) however do not accept the correlation found by Wilson and Watkins (1967) and in their reanalysis of the same data find no significant correlation when field relations and silicate petrography are taken into account.

Watkins and Haggerty (1968) later examined the magnetic polarity and oxidation state of over 550 specimens in single lavas and dikes in Eastern Iceland. They found no correlation between polarity and the oxidation state in the dikes, although there was a strong correlation between the percentage of reversed polarity and higher oxidation in the lavas. In a later paper Ade-Hall

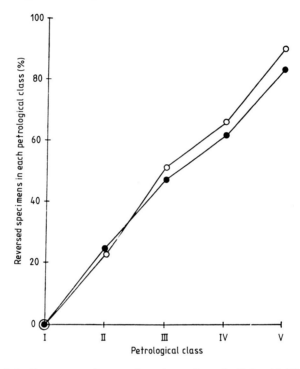

Figure 2.4   Percentage of reversed specimens from the Columbia Plateau basalts for different oxidation states. Petrological Classes I to V show increased oxidation. Open circles: percentage of reversed specimens in each oxidation class, solid circles: percentage of reversed specimens with increasing oxidation state (after Wilson and Watkins 1967).

and Watkins (1970) found no correlation between opaque petrological parameters and polarity in specimens of Canary Island lava flows of Miocene to Pliocene age. The proposed correlation thus does not seem to hold on a worldwide scale and a satisfactory explanation of those positive correlations that have been found has not yet been given. Domen (1969) has reported consistent differences between the demagnetisation curves and Curie points of normally and reversely magnetised basalts from Kawajiri, Japan, and shown the existence of self-reversals in the reversed samples by comparison of thermal demagnetisation curves (the explanation of these effects is not clear).

Kristjansson and McDougall (1982) have carried out a detailed study of the late Tertiary magnetic field in Iceland using more than 2400 lava flows, and found that magnetic moment versus latitude curves were similar for normal and reverse polarities. From this, it would seem that both the actual mean strength of the dipole field and the rock magnetic properties of the basalts are independent of polarity. The supposed positive correlation between oxidation state and reverse magnetic polarity in rocks cannot have applied in this case (as highly oxidised lavas are generally more intensely magnetised and more stable than others), unless the reverse primary magnetic fields were weaker than normal fields by a factor precisely cancelling the effects of the oxidation. Kristjansson and McDougall suggest that it is likely that much of the original sample material on which this correlation was based simply had not received adequate demagnetisation treatment to rid the less oxidised samples of Brunhes age viscous magnetisation.

If field reversal is the cause of reversely magnetised rocks and the correlation between polarity and oxidation state is real, then it implies that there is a statistical correlation between the polarity of the field which originates in the core and the oxidation state of some lava flows which originate in the upper mantle. Since only small variations in core motions are required to reverse the polarity of the dipole field, these could be caused by changes in the boundary conditions at the MCB. Such boundary conditions could be affected by motions in the lower mantle and it is thus not implausible that reversals are correlated to some extent with other phenomena that may be affected by motions in the mantle (Hide 1967). In this connection Hide and Malin (1970) found a correlation between the Earth's gravitational field and the non-dipole geomagnetic field, which they attributed to the effect of topography at the MCB. Irving (1966) has also suggested that the magnetic field would reverse frequently during times of active convection and tectonism (e.g. in the late Cenozoic).

If the origin of reversals is one of the instantaneous self-reversal mechanisms, then normal and reversely magnetised rocks should be randomly distributed throughout a group of rocks of different ages. If reversals are due to one of the time-dependent self-reversal mechanisms, reversals should be increasingly abundant in older rocks. If, on the other hand, reversals are due to geomagnetic field reversals, normal and reversely magnetised groups of rocks should be exactly the same age over the entire Earth; and unless it so happened that the Earth's field suffered more reversals in the past, the proportion of reversed magnetisations should not be greater among older rocks. Almost all rocks baked by igneous bodies agree in polarity with that of the baking rock, irrespective of their compositions (Wilson 1962, 1966b). Self-reversal would produce baking and baked rock pairs of opposite polarity because the supposed self-reversal property is unlikely to be common to each member of a given pair. Furthermore, in cases where a wide variety of rock types in a single locality all possess reversed polarity, it is extremely unlikely that they all contain the self-reversing property. Again all rocks of a given age, at least within the past 4 Myr, have the same polarity irrespective of rock type, composition or location. Self-reversal would have produced a random distribution of reversely magnetised rocks throughout this period. Moreover, the fact that about half of all the rocks examined are reversely magnetised is just what would be expected from field reversal—self-reversal would produce this proportion only by coincidence.

Examples have been found of time-ordered sequences of lavas that apparently record the change of the ancient magnetic field from one polarity to the other, with directionally intermediate steps. Self-reversal cannot explain a gradual swing of direction of magnetisation from one lava flow to another. A more reasonable explanation is that these examples record field reversal actually taking place. Finally, it is found that the range and mean value of the ancient field intensity that is deduced from normally magnetised rocks is the same as that deduced from reversely magnetised rocks. It is unlikely that a self-reversal mechanism would produce the same values as occur in normally magnetised rocks and this suggests, again, that the Earth's field has reversed. Thus, although it has been established that self-reversal does occur in some rocks, the total evidence is overwhelmingly in favour of field reversal in the great majority of cases.

References

Ade-Hall J and Watkins N D 1970 Absence of correlation between opaque petrology and natural remanence polarity in Canary Island lavas *Geophys. J.* **19** 351

Balsey J R and Buddington A F 1954 Correlation of reverse remanent magnetism and negative anomalies with certain minerals *J. Geomag. Geoelec.* **6** 176

——1958 Iron–titanium oxide minerals, rocks and aeromagnetic anomalies of the Adirondack area, New York *Econ. Geol.* **53** 777

Bhimasankaram V L S 1964 Partial self-reversal in pyrrhotite *Nature* **202** 478

Burek P J 1970 Magnetic reversals; their application to stratigraphical problems *Bull. Am. Assoc. Petrol. Geol.* **54** 1120

Carmichael C M 1959 Remanent magnetism of the Allard Lake ilmenites *Nature* **183** 1239

——1961 The magnetic properties of ilmenite–haematite crystals *Proc. R. Soc.* A **263** 508

Channell J E T 1978 Dual magnetic polarity measured in a single bed of Cretaceous pelagic limestones from Sicily *Z. Geophys.* **44** 613

Creer K M, Hedley I G and O'Reilly W 1975 Magnetic oxides in geomagnetism in *Magnetic Oxides* ed D J Craik (New York: Wiley)

Creer K M, Mitchell J G and Valencio D J 1971 Evidence for a normal geomagnetic field polarity event at 263 ± 5 Myr BP within the late Palaeozoic reversal interval *Nature Phys. Sci.* **233** 87

Creer K M and Petersen N 1969 Thermochemical magnetisation in basalts *J. Geophys.* **35** 501

Creer K M, Hedley I G and O'Reilly W 1975 Magnetic oxides in geomagnetism magnetisation and anisotropy of viscous magnetisation in basalts *Geophys. J.* **21** 471

Denham C R 1981 Viscous demagnetisation and the longevity of palaeomagnetic polarity messages *Geophys. Res. Lett.* **8** 137

Domen H 1969 An experimental study on the unstable natural remanent magnetisation of rocks as a palaeomagnetic fossil *Bull. Fac. Educ. Yamaguchi Univ.* **18**(2) 1

Elston D P and Purucker M 1979 Detrital magnetisation in red beds of the Moenkopi Formation *J. Geophys. Res.* **84** 1653

Everitt C W F 1962 Self reversal in a shale containing pyrrhotite *Phil. Mag.* **7** 831

Fujiwara Y and Ohtake T 1975 Palaeomagnetism of Late Cretaceous alkaline rocks in the Nemuro Peninsula, Hokkaido, NE Japan *J. Geomag. Geoelec.* **26** 549

Fuller M D 1970 Geophysical aspects of palaeomagnetism *Crit. Rev. Solid State Phys.* **1** 137

Gorter E W and Schulkes J A 1953 Reversal of spontaneous magnetisation as a function of temperature in Li Fe Cr spinels *Phys. Rev.* **90** 487

Graham J W 1949 The stability and significance of magnetism in sedimentary rocks *J. Geophys. Res.* **54** 131

Havard A D and Lewis M 1965 Reversed partial thermo-magnetic remanence in natural and synthetic titano–magnetites *Geophys. J.* **10** 59

Heller F 1980 Self-reversal of natural remanent magnetisation in the Olby–Laschamp lavas *Nature* **284** 334

Heller F and Egloff R 1974 Multiple reversals of natural remanent magnetisation in a granite–aplite dyke of the Bergell Massif (Switzerland) *J. Geomag. Geoelec.* **26** 499

Heller F, Markert H and Schmidbauer E 1979 Partial self-reversal of natural remanent magnetisation of an historical lava flow of Mt Etna (Sicily) *J. Geophys.* **45** 235

Heller F and Petersen N 1982a The Laschamp excursion *Phil. Trans. R. Soc.* A **306** 169

——1982b Self-reversal explanation for the Laschamp–Olby geomagnetic field excursion *Phys. Earth Planet. Int.* **30** 358

Helsley C E and Steiner M B 1974 Palaeomagnetism of the lower Triassic Moenkopi Formation *Geol. Soc. Am. Bull.* **85** 457

Hide R 1967 Motions of the Earth's core and mantle and variations of the main geomagnetic field *Science* **157** 55

Hide R and Malin S R C 1970 Novel correlations between global features of the Earth's gravitational and magnetic fields *Nature* **225** 605

Hoffman K A 1982 Partial self-reversal in basalts containing mildly low-temperature oxidised titanomagnetite *Phys. Earth Planet. Int.* **30** 357

Irving E 1964 *Palaeomagnetism and its Application to Geological and Geophysical Problems* (New York: Wiley)

——1966 Palaeomagnetism of some Carboniferous rocks from New South Wales and its relation to geological events *J. Geophys. Res.* **71** 6025

Ishikawa Y and Syono Y 1963 Order–disorder transformation and reverse thermo-remanent magnetism in the Fe Ti O_3–Fe_2O_3 system *J. Phys. Chem. Solids* **24** 517

Kennedy L P 1981 Self-reversed thermoremanent magnetisation in a late Brunhes dacite pumice *J. Geomag. Geoelec.* **33** 429

Kristjansson J and McDougall I 1982 Some aspects of the late Tertiary geomagnetic field in Iceland *Geophys. J.* **68** 273

Kropaček V 1968 Self-reversal of spontaneous magnetisation of natural cassiterite *Stud. Geophys. Geod. Ceskoslov. Accad. Ved.* **12** 108

Larson E E and Strangway D W 1966 Magnetic polarity and igneous petrology *Nature* **212** 756

——1968 Discussion: 'Correlation of petrology and natural magnetic polarity in Columbia Plateau basalts by R L Wilson and N D Watkins' *Geophys. J.* **15** 437

Larson E E, Walker T R, Patterson P E, Hoblitt R P and Rosenbaum I G 1982 Palaeomagnetism of the Moenkopi Formation, Colorado Plateau; basis for long term model of acquisition of chemical remanent magnetism in red beds *J. Geophys. Res.* **87** 1081

Lawson C A, Nord G L Jr, Dowty E and Hargraves R B 1981 Antiphase domains and reverse thermoremanent magnetism in ilmenite–hematite materials *Science* **213** 1372

McElhinny M W 1969 The palaeomagnetism of the Permian of south-east Australia and its significance regarding the problem of intercontinental correlation *Spec. Publ. Geol. Soc. Aust.* **2** 61

Nagata T, Akimoto S and Uyeda S 1953 Self-reversal of thermoremanent magnetism of igneous rocks (III) *J. Geomag. Geoelec.* **5** 168

Nagata T and Uyeda S 1959 Exchange interaction as a cause of reverse thermo-remanent magnetism *Nature* **184** 890

Nagata T, Uyeda S and Akimoto S 1952 Self-reversal of thermoremanent magnetism of igneous rocks *J. Geomag. Geoelec.* **4** 22

Néel L 1949 Théorie du trainage magnétique des ferromagnétiques au grains fins avec applications aux terres cuites *Ann. Geophys.* **5** 99

——1951 L'inversion de l'aimantation permanente des roches *Ann. Geophys.* **7** 90

——1955 Some theoretical aspects of rock magnetism *Phil. Mag. Suppl. Adv. Phys.* **4** 191

Ozima M and Ozima M 1967 Self-reversal of remanent magnetisation in some dredged submarine basalts *Earth Planet. Sci. Lett.* **3** 213

Petersen N and Bleil U 1973 Self-reversal of remanent magnetism in synthetic titanomagnetites *Z. Geophys.* **39** 965

Petherbridge J 1977 A magnetic coupling occurring in partial self-reversal of magnetism and its association with increased magnetic viscosity in basalts *Geophys. J.* **50** 395

Purucker M E, Shoemaker E M and Elston D P 1980 Early acquisition of characteristic magnetisation in red beds of the Moenkopi Formation (Triassic), Gray Mountain, Arizona *J. Geophys. Res.* **85** 997

Robertson W A 1963 Palaeomagnetism of some Mesozoic intrusives and tuffs from eastern Australia *J. Geophys. Res.* **68** 2299

Ryall P J C and Ade-Hall J M 1975 Laboratory induced self-reversal of thermo-remanent magnetisation in pillow basalts *Nature* **257** 117

Sasajima S and Nishida J 1974 On the self-reversal of TRM in a highly oxidised submarine basalt *Rock Mag. Palaeogeophys.* **2** 5

Schult A 1968 Self-reversal of magnetisation and chemical composition of titanomag-netites in basalts *Earth Planet. Sci. Lett.* **4** 57

——1976 Self-reversal above room temperature due to N-type magnetisation in basalt *J. Geophys.* **42** 81

Stacey F D and Banerjee S K 1974 *The physical principles of rock magnetism* (Amsterdam: Elsevier)

Tucker P 1980 A grain mobility model of post-depositional realignment *Geophys. J.* **63** 149

Tucker P and O'Reilly W 1980 Reversed thermoremanent magnetisation in synthetic titanomagnetites as a consequence of high temperature oxidation *J. Geomag. Geoelec.* **32** 341

Uyeda S 1955 Magnetic interactions between ferromagnetic minerals contained in rocks *J. Geomag. Geoelec.* **7** 9

——1958 Thermoremanent magnetism as a medium of palaeomagnetism, with special reference to reverse thermoremanent magnetism *Japan J. Geophys.* **2** 1

Walker T R, Larson E E and Hoblitt R P 1981 Nature and origin of hematite in the Moenkopi Formation (Triassic), Colorado Plateau: a contribution to the origin of magnetism in red beds *J. Geophys. Res.* **86** 317

Watkins N D and Haggerty S E 1968 Oxidation and polarity variations in Icelandic lavas and dikes *Geophys. J.* **15** 305

Westcott-Lewis M F and Parry L G 1971 Thermoremanence in synthetic rhombohedral iron–titanium oxides *Aust. J. Phys.* **24** 735

Wilson R L 1962 The palaeomagnetism of baked contact rocks and reversals of the Earth's magnetic field *Geophys. J.* **7** 194

——1966a Further correlations between the petrology and the natural magnetic polarity of basalts *Geophys. J.* **10** 413

——1966b Palaeomagnetism and rock magnetism *Earth Sci. Rev.* **1** 175

Wilson R L and Watkins N D 1967 Correlation of petrology and natural magnetic polarity in Columbia Plateau basalts *Geophys. J.* **12** 405

3 The morphology of geomagnetic reversals

3.1 Introduction

During a polarity change, the direction of the geomagnetic field swings through about 180°, the virtual geomagnetic poles[†] following widely different paths for different transitions. This could be interpreted as the result of a decrease in the main dipole field so that the observed field is dominated by the non-dipole component. Alternatively, the field could just tip over being still mainly controlled by the dipole term with independently varying axial and equatorial components. In the first model the pole paths for transitions would not be expected to be the same whether recorded at different places at the same time or the same place at different times because of variations in the non-dipole field. The second model suggests that there could be strong similarities between pole paths for the same transition observed anywhere on the Earth. This question will be discussed in more detail later.

Four major normal and reversed sequences have been found during the past 3.6 Myr. These major groupings have been called geomagnetic polarity epochs and have been named by Cox *et al* (1964) after people who have made significant contributions to geomagnetism. Superimposed on these polarity

---

† The virtual geomagnetic pole (VGP) is defined as the pole of the dipolar field which gives the observed direction of magnetisation at the site under consideration. It is calculated from any spot reading of the field direction, the word 'virtual' meaning that no implication about the position of an average dipole is being made. To compare data from different sampling sites at different latitudes it is convenient to calculate the equivalent dipole moment which would have produced the measured intensity at the calculated palaeolatitude (assuming a dipolar field) of the sample. Such a calculated dipole moment is called a virtual dipole moment (VDM) and has the advantage that no scatter is introduced by any wobble of the main dipole, since the determined magnetic palaeolatitude is independent of the orientation of the dipole relative to the Earth's rotational axis. However it is not possible to determine that portion of the observed intensity (and therefore of the VDM) which arises from non-dipole components. Hence a true dipole moment (TDM) cannot be obtained directly from palaeointensity studies.

epochs are brief fluctuations in magnetic polarity with a duration that is an order of magnitude shorter. These have been called polarity events and have been named after the localities where they were first recognised (see figure 3.1). The reality of such a distinction has since been questioned. The question of terminology is discussed in chapter 6 on magnetostratigraphy. In this chapter the older terms 'epoch' and 'event' will be retained. Later Opdyke *et al* (1966) found a polarity record in deep-sea sediments going back 3.6 Myr in which the pattern of reversals was remarkably similar to that found in igneous rocks on land (see figure 3.2). More recently cases have been found of more irregular 'excursions' of the poles, or 'aborted' reversals—these will be discussed in chapter 4.

The timescale of reversals during the last 3.6 Myr was originally determined by radiometric dating and palaeomagnetic measurements on volcanic rocks. It was later extended to 11 Myr by Vine (1966) using the palaeomagnetic record of reversals in the oceanic crust adjacent to the East Pacific Rise. It was further extended to 76 Myr by Heirtzler *et al* (1968) using similar data from the South Atlantic. In both studies, the previously determined timescale for the past 3.5 Myr was used to calibrate the rate at which new sea floor was being formed. The rest of the timescale was then determined from marine magnetic profiles by extrapolation. McDougall *et al* (1977) were able to extend the polarity timescale based on radiometrically dated subaerial igneous rocks back another 2 Myr to approximately 6.5 Myr. They used a long stratigraphi-cally controlled sequence of rocks—more than four hundred successive lavas in Borarfyördur, western Iceland, with a total thickness of more than 3500 m. They found two further normal events—very short ones at 6.43 Myr and 6.12 Myr BP. Apparently these have not been detected previously either in marine magnetic anomalies or in deep-sea cores. Harrison *et al* (1979) used thick piles of lava flows in Iceland to obtain independent evidence for the ages of reversals back to 13.0 Myr, and found that in general the assumption of a constant rate of sea-floor spreading is valid. The timescale of reversals is discussed in more detail in chapter 6.

In recent years several attempts have been made to obtain a more detailed timescale by stacking profiles to reduce incoherent noise. This has revealed a fine structure of previously undetected short polarity intervals (Blakely and Cox 1972, Blakely 1974). For the time interval from 22.7 to 7.3 Myr BP Blakely found eighteen new reversals, increasing the average frequency of reversals by 50%. It is very difficult to resolve polarity intervals shorter than about 20 000 yr. New oceanic crust does not form at the rises by a smooth, steady-state process—rather it forms by a sequence of discrete volcanic eruptions separated in space by several kilometres and in time by about 14 000 yr. As a result, even when a magnetometer is towed near the bottom to achieve better spatial resolution, there is too much noise of geological origin to permit polarity intervals shorter than about 20 000 yr to be resolved. Attempts to identify polarity intervals as short as 20 000 yr using the palaeomagnetic

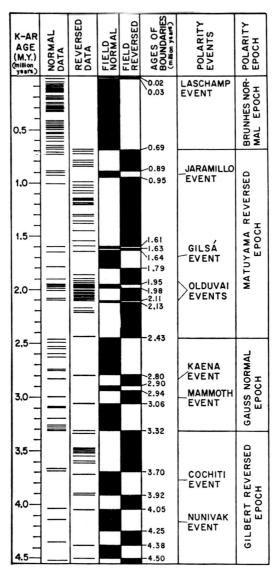

Figure 3.1 Timescale for geomagnetic reversals. Each short horizontal line shows the age as determined by potassium–argon dating and the magnetic polarity (normal or reversed) of one volcanic cooling unit. Normal polarity intervals are shown by the solid portions of the 'field normal' column, and reversed polarity intervals by the solid portions of the 'field reversed' column. The duration of events is based in part on palaeomagnetic data from sediments and profiles (after Cox 1969).

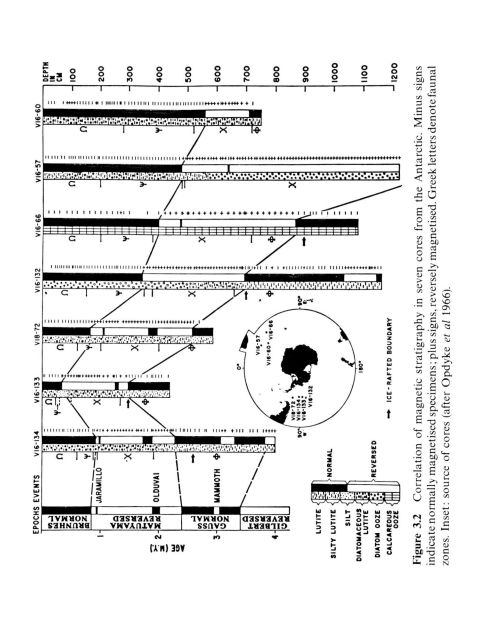

Figure 3.2 Correlation of magnetic stratigraphy in seven cores from the Antarctic. Minus signs indicate normally magnetised specimens; plus signs, reversely magnetised. Greek letters denote faunal zones. Inset: source of cores (after Opdyke *et al* 1966).

record of sediments accumulated on the sea floor has also, in general, proved unsuccessful. Data obtained from different cores of the same age are often inconsistent—probably because of bioturbation of the sediments. It is also often difficult to know when a hiatus in sedimentation has occurred. Finally, in the detection of polarity intervals on land using radiometric dating, the analytic precision of the various chemical analyses involved is about 3%. Thus, even for the most recent reversal about 0.7 Myr ago, this introduces an error of approximately 21 000 yr. Beyond about 3.5 Myr, the precision of the K–Ar method begins to approach the average duration of individual polarity events and the determination of boundary ages becomes increasingly ambiguous. Also K–Ar ages of oceanic basalts are subject to large errors because of hydrothermal alteration and weathering. It is for this reason that the older portions of the polarity timescale have been determined by extrapolating marine magnetic anomaly data.

Palaeomagnetic results cannot always be interpreted unambiguously. For example much work has been carried out in recent years on Keweenawan lavas and intrusives around Lake Superior, Canada. Most results indicate an older period of reverse magnetisation giving way to a younger normal period. However in a sequence from Mamainse Point, Palmer (1970) found a triple reverse sequence ($R{\rightarrow}N{\rightarrow}R{\rightarrow}N$), and suggested that this was due to a strike fault causing repetition of the single reversal sequence found elsewhere. Robertson later (1973) confirmed Palmer's findings and supported his hypothesis. However, Massey (1979) pointed out that there is no geological evidence for strike faulting and carried out major and trace element geochemical studies on the lavas in the critical region and found that the geochemical data also argued strongly against any faulting.

In some cases palaeomagnetic polarities are asymmetric, i.e. normal and reversed directions are not exactly 180° apart. This is true of late Precambrian Keweenawan rocks ($\sim$1200–1000 Myr old) around Lake Superior. Here the inclinations of reversely magnetised rocks are consistently much steeper (upward) than those of normally (downward) magnetised rocks. Four models have been suggested to explain these asymmetries (Nevanlinna and Pesonen 1983):

  (i)   a secondary overprint that when superimposed upon both normal and reversed remanence vectors changed the originally symmetric reversal into an asymmetric one,
  (ii) apparent polar wandering,
  (iii) a single offset dipole (Wilson 1972),
  (iv) a coaxial two-dipole model.

Nevanlinna and Pesonen showed that only (iv) can explain all the data. This model consists of a geocentric dipole and an offset dipole located at the MCB. The offset dipole simulates the long-term zonal average of the spherical harmonic non-dipole field. If the geocentric dipole reverses whilst the offset dipole maintains a constant polarity, asymmetric polarities result. This model suggests that inclination asymmetries are due to a different non-dipole ratio during normal and reversed polarity and satisfactorily explains the three asymmetric reversals at Mamainse Point referred to above.

Pesonen and Halls (1983) have also examined the palaeomagnetic field in late Precambrian Keweenawan rocks and found that on the average the palaeointensities of reversely magnetised rocks are 40% higher than those with normal polarity. They showed that this result cannot be explained by any difference in remanence characteristics, grain size or cooling rate (dyke width) between the rocks of the two polarity groups. The differences, however, disappear when the data are reduced to the palaeoequator in accordance with a geocentric axial dipole field. The palaeointensity results are thus consistent with the suggestion that the reversal asymmetry is caused by the North American plate occupying a lower palaeolatitude when the field had normal polarity. However, such an explanation becomes invalid if three successive asymmetric reversals occurred at Mamainse Point. In this case, the two dipole field configuration of Nevanlinna and Pesonen (1983) best explains the data.

McElhinny (1971) has analysed 1231 palaeomagnetic results for different geologic periods in the Phanerozoic in order to find the proportion of normal to reversed polarities. His results are given in figure 3.3 which shows the percentage of normal polarity plotted as a function of time and is a measure of the proportion of the time that the geomagnetic field has normal polarity. For the Upper Tertiary 50% of the measurements have normal polarity, whilst in

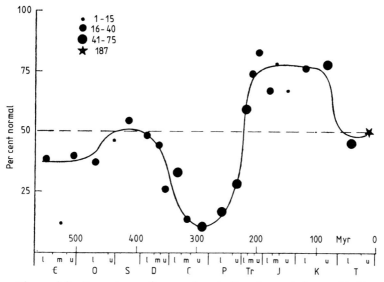

Figure 3.3 Percentage of normal polarity observed in worldwide palaeomagnetic investigations for the Phanerozoic. The different symbols refer to the number of observations (after McElhinny 1971).

the Lower Tertiary 45% are normal. The geomagnetic field in the late Tertiary in Iceland has been examined in great detail by Kristjansson and McDougall (1982). Samples from north-west Iceland contained 594 flows of normal polarity and 523 of reversed polarity. Wilson *et al* (1972) had reported earlier that their eastern Iceland collection contained 576 normal and 450 reversed flows and concluded that over the past 13 Myr the normal state has been more stable than the reversed state. It is thus clear that one cannot make dogmatic statements from one survey about the proportion of time that the geomagnetic field has a particular polarity. The latest results of Kristjansson and McDougall confirm the earlier conclusion of McElhinny (1971) and Einarsson and Sigurgeirsson (1955) that in the late Tertiary either polarity state was equally probable. However, figure 3.3 shows that there is a marked difference between the Lower and Middle Devonian, which have roughly equal occurrences of normal and reversed data, and the Upper Devonian which is largely reversed and marks the beginning of a predominantly reversed era lasting until the close of the Palaeozoic. For the mid-Carboniferous to the Lower Permian only a few results show normal polarity. Apart from the Lower Triassic, where frequent reversals have been observed in several formations, the Mesozoic has predominantly normal polarity ($\sim 75\%$); measurements in the Upper Triassic indicate that as much as 83% of the time, the field was normal. A more recent analysis by Irving and Pullaiah (1976) confirms this general pattern. The question of polarity 'bias' is discussed further in §5.4.

3.2 Field intensity during a polarity transition

A polarity transition takes place so quickly (on a geological timescale) that it is difficult to find rocks that have preserved a complete and accurate record. Good intensity estimates may be obtained for volcanic rocks but suffer from the fact that there is little chronological control. On the other hand, sedimentary rocks give reasonably good chronological control, but sedimentation rates are often too slow to allow detailed resolution of intermediate fields. Moreover, an initially complete and accurate palaeomagnetic record in sediments may be obliterated or altered by chemical diagenesis and bioturbation occurring after initial deposition.

Observational evidence on changes in intensity during a reversal is conflicting. Although there seems to be general agreement that there is a marked reduction in intensity associated with the reversal (perhaps to as little as 10% of the initial field on occasion) there is disagreement as to whether the onset of changes in field direction coincides with the reduction in intensity. One of the earliest studies was made by Van Zijl *et al* (1962) on rocks from the Stormberg volcanic sequence in Africa. They found a major reduction in intensity during the reversal with some indication of an intensity increase immediately before and after the actual reversal. Similar results were obtained by Soviet workers using sedimentary sections on land (see e.g. Kaporovich *et al* 1966).

Several workers succeeded later in obtaining a record of the geomagnetic field during a polarity transition (see e.g. Watkins 1967, 1969, Lawley 1970). It appears that during a reversal the intensity of the field first decreases by a factor of three or four for several thousand years while maintaining its direction. The magnetic vector then usually executes several swings of about 30°, before moving along an irregular path to the opposite polarity, the intensity still being reduced, rising to its normal value later. It is not certain whether the field is dipolar during a transition. Moreover there do not seem to be any precursors of a reversal or any indication later that a reversal has occurred. A detailed record of a field reversal has been described by Dunn *et al* (1971) who examined a single, igneous intrusive body (the Tatoosh intrusion in Mount Rainier National Park, Washington) which cooled comparatively slowly from the outside. If such a body cooled slowly enough, a large timespan of the ancient geomagnetic field would be recorded continuously as the Curie point isotherm slowly moved further into the intrusion. This would avoid the discontinuities inherent in lava suites and the poor resolution inherent in sedimentary sequences. It is difficult however, to establish an absolute timescale, since it depends upon calculations of cooling rates based upon parameters which are themselves poorly known. Dunn *et al* found that the field intensity decreased by a factor of ten before any change in field direction occurred, and did not return to normal until after the directional change was completed. The directional change was estimated to have taken 1000–4000 yr, while the intensity change took 10 000 yr.

Contradictory results on the behaviour of the geomagnetic field during a reversal were obtained by Opdyke *et al* (1973) from measurements made on a high deposition rate deep-sea core. The core is reversely magnetised to a depth of 460 cm, normal from 460 to 940 cm, followed by a long unbroken stretch of reversed magnetisation to a depth of 2250 cm where a further polarity change occurs. The rest of the core is normally magnetised. The authors identified the normal interval between 460 and 940 cm as the Jaramillo event. The duration of the Jaramillo event has been estimated by Opdyke (1969) to be 56 000 yr, making the rate of sedimentation during the event 8.6 cm/1000 yr in this core. Because of this high rate of deposition, intermediate directions of magnetisation were observed within each transition as well as a sharp decrease in intensity.

Details of the behaviour of the magnetic field at the lower Jaramillo reversal are shown in figure 3.4. It can be seen that the pronounced drop in the intensity of magnetisation is coincident with the onset of the direction changes. The time taken for both the intensity and direction of the field to reverse is approximately 4600 yr based on a sedimentation rate of 8.6 cm/1000 yr. Figure 3.4 also shows that there are three cycles of intensity changes with periods of

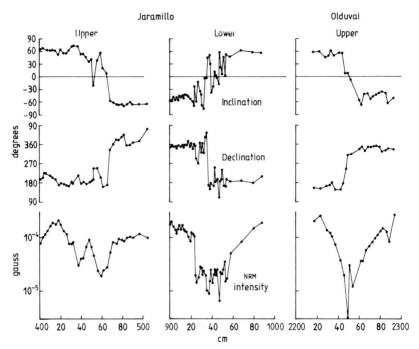

Figure 3.4 Details of the inclination, declination and intensity of magnetically cleaned remanence across the upper and lower Jaramillo, and the upper Olduvai polarity transitions (after Opdyke *et al* 1973).

approximately 1400 yr during the time of the polarity change. The declination and inclination variations show a similar periodicity. The period of these movements is close to that usually associated with the secular variation of the non-dipole field—supporting the hypothesis that during the transition the dipole field is weak, being of comparable intensity to that of the non-dipole field. Opdyke *et al* also concluded that the decreases in intensity were not caused by variations in the magnetic mineralogy of the sediments but reflect actual decreases in intensity of the Earth's field.

Kawai *et al* (1976) studied the Brunhes–Matuyama transition using thin sections of a core from the Melanesia Basin. They found the onset of directional change to be coincident with a sharp drop in intensity. They suggested that this has not been clearly seen before because with average sedimentation rates, the usual length of core samples is too large to reveal fine structure: the core sampled by Opdyke *et al* (1973) had a relatively high sedimentation rate. Bingham and Evans (1975) also found the duration of the reduction in intensity to be comparable to that for the change in direction in a reversal seen in the Stark formation, Great Slave Lake, Canada in Precambrian red siltstones (approximate age 1650–1800 Myr). They found, however, some suggestion of very rapid variations in intensity immediately before and after the change in direction. Nevertheless in the majority of cases studied, the intensity change appears to be longer than the change in direction (see e.g. Dunn *et al* 1971, Baag and Helsley 1974a, b, Hillhouse and Cox 1976, Denham *et al* 1977). Dodson *et al* (1978) in a detailed study of a reversal recorded in the Tatoosh intrusion in Mount Rainier National Park found that the intensity decreased substantially before the onset of directional changes, in agreement with the earlier work of Dunn *et al* (1971). The change in declination was gradual at first but the inclination changed rapidly initially. Moreover the dispersion of directions, as measured by the Fisher precision parameter K, increased markedly during the reversal, reaching a maximum at the time of minimum intensity of magnetisation (see figure 3.5).

Herrero-Bervera and Helsley (1983) have analysed the palaeomagnetic record in samples collected from the Red Peak Member of the Chugwater Formation of Triassic age in central Wyoming. They found that a slow change in intensity without change in direction begins long before the marked intensity decrease associated with the change in direction. The major decrease

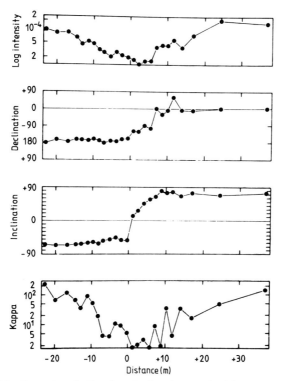

Figure 3.5 Intensity, declination, inclination, and Fisher parameter K plotted against perpendicular distance from reversal isotherm plane (after Dodson *et al* 1978).

in intensity appears to be coincident with the directional change and recovers shortly after the directional change is complete. Thus, although the major portion of the intensity change took the same time as the directional change, the total time involved in the transition may be as much as three times that associated with the period of rapid directional change. Their results indicate that the dipole field 'collapsed' rather than 'flipped over' during the reversal.

It is noteworthy that the polarity change seen by Bingham and Evans shows that reversals have been a feature of the geomagnetic field for at least 2000 Myr. Their transition lasted about 10^4–10^5 yr which is considerably longer than that for more recent reversals. Ito (1970), in a study of volcanic rocks of Pliocene to Miocene age from south-western Japan also found that the time of transition was longer (10^4–10^5 yr). Some interesting results have been found by Kawai *et al* (1977) on a core obtained from the Melanesia Basin. They found a number of cases when the intensity of the magnetic field vanished. The youngest instance was at the Brunhes–Matuyama boundary. Kawai and Nakajima (1975) had earlier estimated that the field was then reduced by two orders of magnitude for about 10 000 yr. The next instance occurred at the lower Jaramillo boundary when the field vanished for more than 13 000 yr. They found yet another case of zero field from about 1.05–1.07 Myr ago. In this case the field recovered in the same direction and thus the 'excursion' remained unnoticed. The oldest example corresponds to what has been called the Reunion event—in this case also the field vanished and recovered in the same direction.

Bol'shakov and Solodovnikov (1980) analysed all world data on intensities measured on continental extrusive and baked rocks (excluding oceanic basalts), ranging in age from the Early Triassic to the Brunhes epoch. They concluded that the average values of the intensity were the same for both normal and reversed fields from the Pliocene to the Late Quaternary, and moreover that for the last 250 Myr the Earth's magnetic moment has remained essentially constant. Kristjansson and McDougall (1982) examined the intensity of the geomagnetic field as recorded in late Tertiary lava flows in Iceland and confirmed that there was no significant difference between normal and reversed flows—see figure 3.6 where intensities are plotted against VGP latitude. Shaw *et al* (1982) used 199 cores from a sequence of Icelandic lavas to determine 68 intensity values of the magnetic field ranging in age from 2 to 6 Myr (see figure 3.7). They found large and rapid changes covering about an order of magnitude. The variation does not follow a normal distribution and there is a tendency for positive swings in magnitude to be both larger and sharper than negative swings. The data also suggest that low field values are more likely to occur during reversed periods. The separate averages of the normal and reversed data are not significantly different, but there is some suggestion of asymmetry in the normal and reversed field states in that the mode, median and quartiles of the reversed data are lower than those of the normal data. As discussed in §2.3, Watkins and Haggerty (1968) found a strong

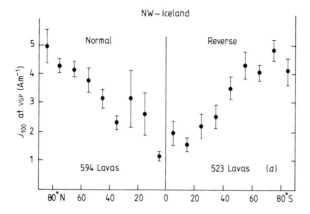

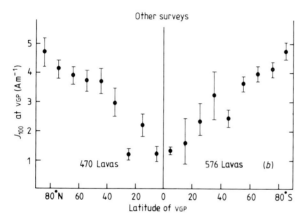

Figure 3.6 (*a*) Arithmetic means of remanence intensities after transformation to virtual geomagnetic pole grouped by 10° increments in VGP latitude. Normal and reversed lavas from north-west Iceland plotted separately. Bars are standard errors. (*b*) Same as (*a*), lavas from other surveys from Iceland (after Kristjansson and McDougall 1982).

correlation between high oxidation states and reversed lavas in Eastern Iceland—Shaw *et al* (1982) suggest that the suspected asymmetry of the magnitude of the palaeofield may be connected in some way with this observation. This question is discussed further in §5.5.

A surprising result has been reported by Shaw (1975). During a well documented transition in a lava flow in western Iceland, he found that the geometric field was large and stable when the VGP was close to the equator (see figure 3.8). This reinforces an earlier, but far weaker, suggestion of Wilson *et al* (1972) of rather strong intermediate dipole moments in an otherwise smooth variation (see figure 3.9). Confirmation of this result by Shaw is stronger since

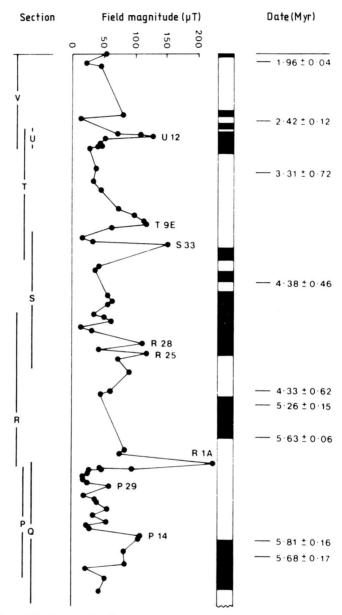

Figure 3.7   Composite diagram of the magnitude of the geomagnetic field with the time axis based on the assumption that on average the lavas are extruded at regular intervals. The time overlap of the sampling sections are shown together with the individual palaeofield magnitude data and the polarity of all measured samples—dark areas represent normal polarity (after Shaw *et al* 1982).

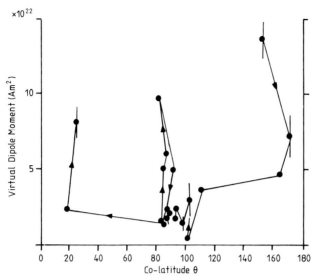

Figure 3.8   Graph of VDM against co-latitude for the R_3 to N_3 transition zone of western Iceland. The error shown is the standard deviation—in most cases it is less than the black circles. Arrows indicate the time progression and show that the large intermediate VDM grow and decay smoothly (after Shaw 1975).

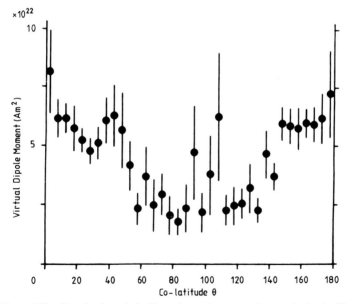

Figure 3.9   Graph of statistically estimated VDM against co-latitude. The dipole moments are averaged over 5° of co-latitude. The error shown is the error of the mean (after Wilson *et al* 1972).

he used only four lavas recording a single transition which is thus much less subject to statistical fluctuations—Wilson *et al* applied statistical analysis to some 3000 specimens from 1500 lavas (Shaw and Wilson 1977). Shaw later (1977) found another case of a strong intermediate palaeofield—in the Lousetown Creek, Nevada transition (see figure 3.10). The fact that the intermediate palaeofield can remain fixed in one direction for a considerable period of time (43 lavas recorded in western Iceland and 34 lavas in Lousetown Creek), and also that it can change in magnitude without changing direction indicates that the Earth's magnetic field may have a third metastable state which occurs much less frequently than the normal and reversed states.

A similar result has been found by Barbetti and McElhinny (1976). Archaeomagnetic studies of prehistoric aboriginal fireplaces along the ancient shore of Lake Mungo, a dried-out lake in south-eastern Australia, revealed

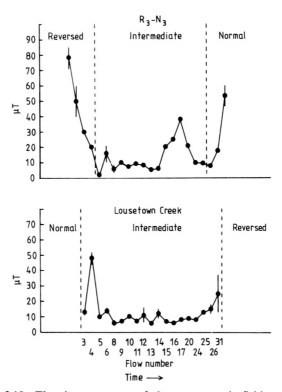

Figure 3.10 The time sequence of the geomagnetic field magnitude during the $R_3 \rightarrow N_3$ transition and the Lousetown Creek transition. The time order is known, the time interval between data points is not known and in some cases may be as long as the extrusion time of several lavas (e.g. between flow 26 and flow 31 no acceptable palaeofield magnitudes were obtained) (after Shaw 1977).

high field strength during an excursion of the geomagnetic field—an increase by a factor of five or six over what appears to be the background field strength outside the excursion and an increase by a factor of three over the present field strength (see figure 3.11). It is possible that the excursion seen at Lake Mungo could be the result of lightning strikes.

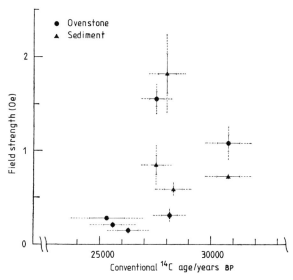

Figure 3.11   Mean ancient field strength at Lake Mungo, Australia. Full (broken) lines show the experimental errors at the one (two) standard error level. Present day field at the locality is 0.6 Oe (after Barbetti and McElhinny 1976).

In a study of the natural remanent magnetisation of cores from Lake Michigan, Dodson *et al* (1977) also found instances of major fluctuations in intensity with little accompanying changes in direction. However, palaeointensities cannot be firmly obtained from sediments and neither Creer *et al* (1976) nor Vitorello and Van der Voo (1977), who also worked on Lake Michigan sediments, reported unduly high intensities.

The absence of directional changes, if further substantiated, suggests that the fluctuation in intensity is in the dipole field, and that the non-dipole field is relatively weak at that time. Such a suggestion raises two questions—what is the cause of the fluctuations in the dipole field, and why is the non-dipole field so weak? In another study of the behaviour of the geomagnetic field during a polarity transition of Upper Miocene age in lavas from Steens Mountain, Oregon, Prevot (1977) found large and rapid changes both in intensity and direction.

3.3 Field direction during a polarity transition

Another unresolved question is whether the magnetic field vector at a given locality tends to move along the same path during successive polarity transitions. If only the non-dipole field were present over much of a transition, the VGP from many reversals would be expected to be randomly scattered through all longitudes. Some analyses have suggested, however, that there may be preferred meridional bands within which most transition pole paths lie. From an analysis of Tertiary polarity transitions, Creer and Ispir (1970) concluded that during a transition a significant equatorial dipolar component remains which can reverse independently of the axial dipole and on a shorter timescale. The movement of the dipole from one hemisphere to the other, corresponding to transition field directions recorded at widely separated sites, follows the same path, passing through the Indian Ocean. Petrova and Rassanova (1976) also found preferred locations for pole paths during reversals of different ages, although a later analysis of the transition field by Vadkovskii *et al* (1980) found no specific trajectories for the movement of the VGP.

Hammond *et al* (1979) analysed the results obtained from two oriented piston cores from the north-west Pacific Ocean (cores K75-01 and K75-02) which recorded the Jaramillo and Olduvai transitions (see figure 3.12). These cores are approximately 120° in longitude away from the Indian Ocean site of Opdyke *et al* (1973). The high accumulation rate of core RC14-14 of Opdyke *et al* allowed them to sample at approximately 200 yr intervals. The lower accumulation rate of the cores of Hammond *et al* meant that the sampling interval during the Jaramillo event was about 1300 and 1000 yr respectively. The resemblance between the VGP paths for cores K75-01, K75-02, and RC14-14 for corresponding transitions, together with the fact that the cores come from widely separated regions, tentatively supports a transition model in which a dipolar field is predominant during each of the polarity reversals studied. On the other hand, successive VGP paths do not coincide and neither Opdyke *et al*'s nor Hammond *et al*'s VGP paths match those determined by Creer and Ispir (1970). Thus while these transition fields seem to retain predominantly dipolar configurations, the dipole (or, possibly, the non-dipole) field may incorporate large-scale magnetic features that can differ between successive reversals.

Hammond *et al* also found a decrease in intensity associated with the change in direction which adds support to the non-dipole transition model. The K75-02 intensity values decreased to about 10% of the values before and after the transition, a level consistent with the present non-dipole field intensity throughout most of the Pacific basin region (Doell and Cox 1972). Moreover the intensity begins to drop in advance of the directional change and recovers only after the directional change is complete—in agreement with what is usually found, although as already noted, Opdyke *et al* found that in

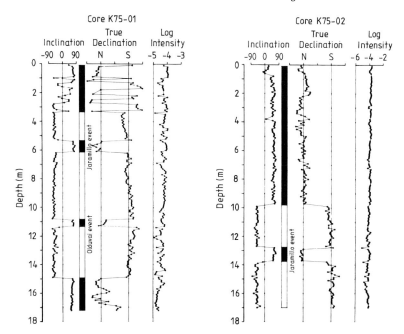

Figure 3.12   Palaeomagnetic data for cores K75-01 and K75-02 from the north-west Pacific Ocean oriented with respect to true north and corrected for corer rotation. Full intervals indicate normal polarity in the core, open intervals indicate reversed polarity (after Hammond *et al* 1979).

their deep-sea core, changes in intensity and direction occurred at the same time.

Steinhauser and Vincenz (1973) investigated the longitudinal and latitudinal distribution of palaeopoles during a polarity transition using data from 23 field reversals ranging in age from Recent to Upper Mesozoic. They found two preferential meridional bands of polarity transition centred on planes through 40°E–140°W and 120°E–60°W. Both these bands are separated by two broad regions without transitional poles situated between 6°E–35°W and 85°E–70°W. The 40°E–140°W meridional band of preferential transitions is in agreement with the results of Creer (1972) who, using twelve polarity changes, determined an average path centred on the plane through 60°E–120°W. It may be significant that the Central Pacific lies within the transition zone between 145°E and 174°W, i.e. within the band between 6°E and 35°W. The secular variation over the Central Pacific has been very small for at least the last million years (Doell and Cox 1971, 1972), suggesting that the non-dipole component of the main field has been very small in this region. Kristjansson and McDougall (1982) in an analysis of the geomagnetic field in the late Tertiary recorded in lava flows from Iceland found that there was no strong preference for the transitional poles to be found in any one particular longitude interval. The most noticeable tendency was for poles to be found in the two regions which are $\pm 90°$ in longitude away from Iceland.

Steinhauser and Vincenz also investigated the latitudinal distribution of transitional palaeopoles. They found a U-shaped distribution, there being a decrease in the number of observed poles with decreasing latitude. They interpreted this result as reflecting an acceleration in the movement of the dipole axis as it approaches the equator—they estimated that it is moving 3.4 times faster in equatorial latitudes than in latitudes around 40°. In contrast the record of a non-dipole field would give a random latitudinal distribution of poles. They further estimated that the dipole moment is reduced by about one order of magnitude for only about 12% of the transition time, while for two-thirds of the time its magnitude is comparatively high—with field intensities considerably greater than the intensity of the non-dipole field. Kristjansson and McDougall (1982) have also studied the latitude distribution of VGP, using as data late Tertiary lava flows from Iceland. Their results (figure 3.13) again illustrate that there is no significant difference between normal and reversed geomagnetic poles. They found that the latitude range where the VGP was most commonly found was around 70°N or S, which is unexpectedly low when compared to the present-day geomagnetic field. The chance that the VGP will be found within 10° of the geographic pole was only 11%. The VGP spent on average 50% of its time below latitude 64.5°, 9% below 35°, and about 2% below 10°.

Opdyke *et al* (1973) traced the movement of the VGP during a reversal. Figure 3.14 shows the movements for the upper and lower Jaramillo transitions. For the lower transition the VGP describes first a clockwise and then an anticlockwise loop in the southern polar regions. It then enters the northern hemisphere, tracing out a clockwise loop before settling down. These three loops are accompanied by the intensity variations described earlier. None of the transition paths of Opdyke *et al* pass through the Indian Ocean as proposed by Creer and Ispir (1970)—the three reversed paths actually occur in different quadrants. The results of Opdyke *et al* thus do not support the model of a toppling dipole: rather they indicate that the intensity of the dipole field drops rapidly to a low but non-zero value, allowing the non-dipole field to predominate. This is reflected in the large looping excursions from the rotation axis seen in the VGP paths (figure 3.14). Since both clockwise and anticlockwise rotations of the VGP path are seen, it would appear that both westward and eastward drifts of the non-dipole field occurred (Skiles 1970). Both eastward and westward drifts of the non-dipole field have also been inferred from the sense of VGP loops associated with the upper Miocene Steens Mountain polarity change (Watkins 1969). Detailed studies of VGP paths during a reversal indicate that such looping is quite common. Gurarii (1981) has studied the Jaramillo event in Western Turkmenia. He found that the VGP followed a complex path with a large number of loops. However the pole positions are confined to a comparatively small region of the Earth's surface—passing

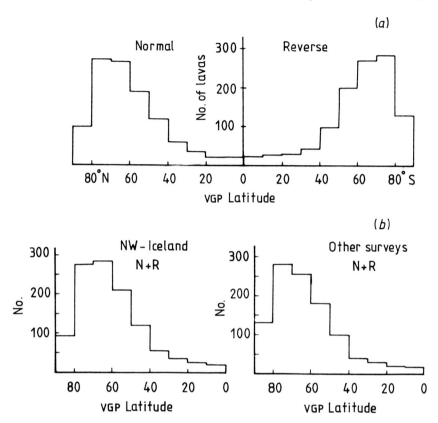

Figure 3.13 (*a*) Histogram of the distribution of VGP positions in latitude. Normal and reversed flows from north-west Iceland plotted separately. (*b*) Normal and reversed flows combined (after Kristjannson and McDougall 1982).

through the South Atlantic and the Eastern part of South America, the North Atlantic and across North America. In their study of the Jaramillo event, Opdyke *et al* (1973) gave two positions of the transitional VGP—one in the south and the other in Japan, i.e. almost 180° in longitude away from the path obtained by Gurarii using data from Western Turkmenia. Gurarii emphasised that the position of just two VGP is insufficient to define the whole pole path during a transition.

Dagley and Lawley (1974) have examined more extensive worldwide data and also found no evidence for a common pole path nor even a preferred sector of the globe. This is in agreement with the analysis of Hillhouse *et al* (1972), who, however, did find a tendency for successive transition paths from the same geographical region to coincide. It must be stressed, however, that the data in general consist of a collection of poles in the two hemispheres with but

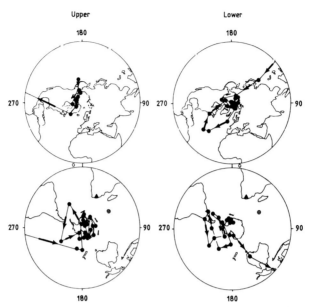

Figure 3.14   Positions of the VGP for upper and lower Jaramillo polarity transitions. VGP are calculated from remanent directions by suitable adjustment of declinations. Core location shown by circled cross. Position of VGP for intermediate direction from Jaramillo Creek, New Mexico, dated at 0.86 Myr (Doell and Dalrymple 1966) shown by solid triangle (after Opdyke *et al* 1973).

little knowledge of the 'path' between them. Dodson *et al* (1978) in an examination of all available paths from North America found that no longitude was strongly preferred, although, as Hoffman (1977) had noted, there appears to be a tendency for several paths to be controlled in part by the site location. In this regard Dodson *et al* found two reversal paths from the same general area but differing in age by about 10 Myr to be essentially the same. In 1977 Hoffman pointed out that for R→N reversals observed at mid latitudes in the northern hemisphere the poles tend to lie predominantly in the hemisphere centred about the site meridian. These he termed near poles. Conversely there is some indication that N→R transitional poles tend to be found in the hemisphere centred about the antimeridian (far-sided poles).

In 1978 Hoffman and Fuller obtained data for the Brunhes–Matuyama reversal from five sites in the northern hemisphere. If the paths are plotted as observed (figure 3.15) it can be seen that there is no correlation between them. However, if the paths are plotted with respect to site longitude (figure 3.16) the paths become confined to the hemisphere centred about the site longitude, i.e. they are near-sided. It is also evident that there tends to be more dispersion of the paths in the second half of the record after the VGP has crossed the equator. Although not always true, the second half of a transition (i.e. after the VGP has

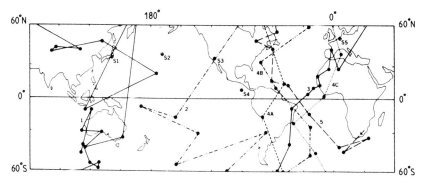

Figure 3.15 VGP paths for Brunhes–Matuyama reversal. Sites 1 Boso peninsula, 2 north central Pacific, 3 Lake Tecopa, 4 east equatorial Pacific, 5 Bruggen (after Fuller *et al* 1979).

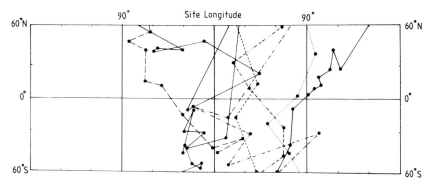

Figure 3.16 VGP paths for Brunhes–Matuyama reversal plotted with respect to site longitude (after Fuller *et al* 1979).

crossed the equator) frequently shows greater scatter in pole positions than the first half. This feature suggests that the field is more strongly controlled by zonal components during the onset of the reversal process—as pointed out previously by Fuller *et al* (1979). Hide (1981) came to the same conclusion on theoretical grounds, suggesting that axisymmetric terms dominate transition fields primarily during the decay phase of the main field. During a reversal there are also patterns of changes in field direction with a 'period' of between one-fifth and one-tenth the time of the total reversal. There is also a tendency for northern hemisphere loops to be clockwise and southern hemisphere loops to be anticlockwise. The mean VGP paths are not traversed smoothly but have periods of rapid motion from one region to another.

Clement *et al* (1982) have also studied the Brunhes–Matuyama transition in three deep-sea sediment cores from the Pacific, two from mid-northern latitudes less than 200 km apart and one from the equatorial Pacific for

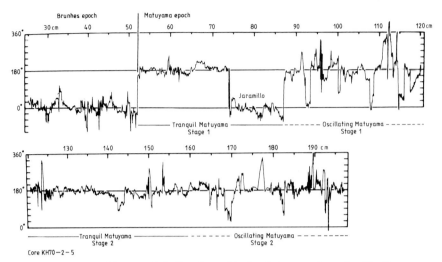

Figure 3.17   Change of declination with depth in core from the North Pacific basin, showing the oscillation of the field during the Matuyama epoch (after Kawai *et al* 1973).

comparison with nearby cores studied by Freed (1977). Estimates of the duration of the transition ranged from 4900 to 8500 yr. It is interesting to note that these estimates are similar to that of 4600 yr given by Opdyke *et al* (1973) for the Lower Jaramillo in spite of there being almost an order of magnitude difference in sedimentation rates. All three transition VGP paths of Clement *et al* are characterised by a portion in which the VGP moves from high southerly latitudes to high northerly latitudes along a longitudinally confined path, lying no more than 90° in longitude away from the site, i.e. the paths tend to be near-sided (Hoffman 1977). However each path contains one or more loops either preceding or following the longitudinal path. Clement *et al* point out that there are some indications in the data that the detailed record during the transition may not be simply related to the geomagnetic field. The VGP paths for the two nearby mid-northern latitudes do not exactly coincide as might be expected considering their geographic proximity. Again the VGP path for their equatorial core is not coincident with those obtained by Freed from nearby cores. Clement *et al* suggest that the magnetic record may be distorted by sedimentological factors such as the effects of burrowing organisms and small hiatuses.

Kawai *et al* (1973) have also examined the time variation of the geomagnetic field during the Matuyama reversal polarity epoch, using a core from the North Pacific basin (see figure 3.17). Between depths of 52–87 cm and 124–142 cm the declination is remarkably constant, except for two sudden changes at the upper and lower Jaramillo transition zone. Kawai *et al* call these periods 'Tranquil Matuyama'. During the rest of the Matuyama epoch there are strong oscillations of the magnetic field, although the total duration of reversed polarity is much longer than that of normal. Kawai *et al* call this stage of the Matuyama epoch 'Oscillating Matuyama'. They speculate that perhaps a new field with opposite polarity existed from the time of the intensity drop at the Brunhes–Matuyama boundary and that perhaps two fields can co-exist for a long time. The 'Tranquil Matuyama' period would then be the consequence of two approximately equal antiparallel poloidal fields, whereas the 'Oscillating Matuyama' regime is the result of an imbalance between the two fields— one fluctuating in intensity relative to the other.

Liddicoat (1981, 1982) has examined the Matuyama–Gauss transition in a core from Searles Valley, California. The transition lasted approximately 2000 yr. About 2000 yr earlier, a short reversal also occurred that lasted about 2000 yr. During it and the main reversal, the relative field intensity decreased by at least 70%. The VGP path for the main reversal is confined to a meridional band 20° wide in the Atlantic Ocean and closely matches the VGP path for the Brunhes–Matuyama transition recorded at Lake Tecopa, California (Hillhouse and Cox 1976). On the other hand, the pole path for the Gauss–Matuyama transition recorded at sites in Russia (Burakov *et al* 1976), about 180° in longitude from Searles Valley, is nearly antipodal at low latitudes to the Searles Valley pole path. This is strong evidence that the transition field was predominantly non-dipolar. This question is discussed further in §3.4.

Williams and Fuller (1981a) obtained an unambiguous *far-sided* VGP path for an R→N reversal recorded in the Agno batholith (a 14–15 Myr old quartz diorite) in Luzon Island, the Philippines. This result is contrary to those obtained from nearly all other R→N northern hemisphere reversals which have near-sided VGP paths and lessens the claim of a general site dependence. In fact Williams and Fuller (1981b) later pointed out that there is no guarantee that each reversal has the same harmonic content. Hoffman (1982) suggested that such an overall characteristic, if it does exist, may be time-dependent or determinable only in a statistical manner.

3.4 Transition fields

There are essentially two classes of models that have been suggested to account for reversals. In the first (the 'standing field' model of Hillhouse and Cox 1976) the main dipole field diminishes while at least parts of the non-dipole field do not. The transition field at any site is thus the sum of a steady non-dipole component and an axial dipole component that changes only in magnitude and sign. Unless the direction of the standing non-dipole field everywhere lies on the north–south vertical plane, there is no site dependence of path longitude. The second model (the 'flooding' model of Hoffman 1977, 1979) is based on the idea that the reversal process floods through the fluid core from a localised zone or point of initiation. The entire dipole source

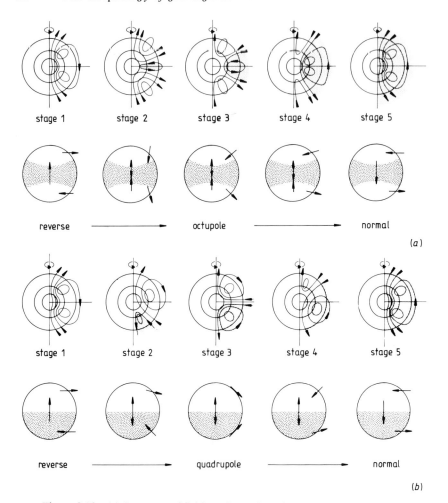

Figure 3.18   (*a*) Sequence of field configurations for R→N reversal with octupolar transition field (after Fuller *et al* 1979). (*b*) Sequence of field configurations for R→N reversal with quadrupolar transition field (after Fuller *et al* 1979).

remains active, but normal and reverse flux is produced simultaneously in different regions. The VGP moves along the great circle defined by the geographical poles and the observer's site. Depending on the sense of the reversal, where in the core source region the reversal process begins and the hemisphere of the observer, the VGP path for the reversal will be 'near-sided' (i.e. the same longitude as the site) or 'far-sided' (180° away). It is not known whether the non-dipole field is predominantly quadrupolar or octupolar. Figure 3.18 shows schematically a sequence of five field configutations for an

R→N reversal. Figure 3.18(a) is for a pure octupolar transition field geometry and 3.18(b) for a pure quadrupolar field. In the upper section of each figure, the lines of force of the poloidal fields are shown—the shaded area represents the region of new polarity and illustrates its growth throughout the reversal. In the lower part of each diagram schematic dipoles are used to illustrate the sources of the observed fields. The shading in the lower section of the diagrams indicates the region of the core in which the reversal was initiated.

It is evident from figure 3.18(b) that the sequence of inclination values for the quadrupolar geometry, whether observed at northern or southern hemisphere sites, includes a vertically downward inclination, implying that the VGP is immediately beneath the site. Hence, each of these records would be near-sided. In contrast, the sequences of inclination values for the octupolar geometry do not both include a $+90°$ inclination. The northern hemisphere site does, and its path will be near-sided; however, the southern hemisphere site has a $-90°$ inclination and would therefore be far-sided, with a VGP antipodal to the site. Fuller *et al* (1979) considered other models in which the reversal is initiated in different areas of the core and which may not depend upon the sense of the reversal. Figure 3.19 lists various possibilities. The examples illustrated in figures 3.18(a) and 3.18(b) are given in the top left-hand quadrants of figures 3.19(a) and 3.19(b) respectively. Both northern and southern hemisphere records of a single reversal are needed to distinguish between quadrupolar and octupolar geometries. Hoffman later developed a model which defines a starting point for the reversal rather than a zone as used in the above quadrupolar and octupolar geometries. This will be discussed in chapter 5. Which of the two models for reversals (the flooding or the standing field model) is correct cannot be resolved by data from only one record of a particular reversal—at least two distant records are required. In 1976 Hillhouse and Cox obtained records of the Brunhes–Matuyama transition from Lake Tecopa, California and from the Boso peninsula, Japan (Niitsuma 1971). The VGP paths were quite different indicating that, at least for this reversal, the transition field was controlled by the non-dipole components.

Hoffman (1981) suggested a test to distinguish between the flooding and standing field models. The former model predicts that, as long as a reversal is initiated in the same zone in the core each time, successive reversals will have transition VGP paths that differ in longitude by $180°$. The latter predicts identical VGP paths for successive reversals. Bogue and Coe (1982) recently obtained data from basalts from Kauai which recorded successive R→N and N→R reversals in the early Pliocene. Their results however were not conclusive. Both reversals are characterised by VGP paths that pass near the site. Although this site dependence is most naturally explained by the zonal flooding model, it is not easy to account for the nearly identical R→N and N→R paths. Their similarity requires the presumed flooding processes for the two reversals to have begun in opposite regions of the core. If the transition field was predominantly octupolar, as is consistent with the very low field

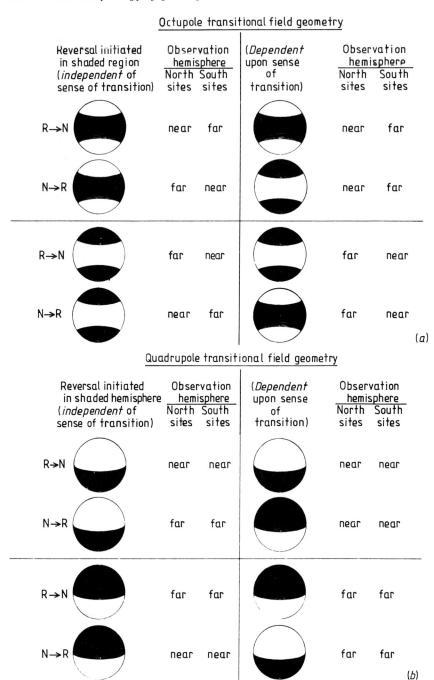

Figure 3.19   (*a*) Predicted VGP paths for octupole transition fields (after Fuller *et al* 1979). (*b*) Predicted VGP paths for quadrupole transition fields (after Fuller *et al* 1979).

intensities that have been inferred from other transition zones, then flooding for the R→N reversal would have begun at the equator. For the N→R reversal, one is then forced to conclude that the reversal was somehow initiated simultaneously at both poles. Quadrupolar transition fields require initiation at the south pole for R→N reversals and at the north pole for N→R reversals. More support for the zonal flooding model has come from back-to-back reversal records of Miocene age in Crete (Valet and Laj 1981) where VGP paths differ by 135°—the VGP path for the N→R reversal follows a longitudinal path almost exactly 180° from the site.

The similarity of the N→R and R→N reversal records from Kauai can easily be explained by the standing field model. Even stronger support for the standing field model comes from California. Volcanic rocks from Clear Lake record intermediate directions from two N→R reversals that occurred several hundred thousand years before the Brunhes–Matuyama transition (Mankinen *et al* 1981). The VGP equivalent to these directions fall right on the Brunhes–Matuyama R→N transition path that Hillhouse and Cox (1976) found at Lake Tecopa. As mentioned in §3.3 additional evidence comes from a detailed record of the Gauss–Matuyama transition in dry lake sediments from Searles Valley, California (Liddicoat 1981, 1982). This reversal, 1.6 Myr earlier and N→R, also repeats the Lake Tecopa path. These three VGP paths are not strongly near- or far-sided, the VGP passing almost 90° to the east of the site. If the standing field controls the transition field, the ages of these sites in California imply that the standing field remained unchanged for several million years.

Valet *et al* (1983) have found two R→N reversals at different stratigraphic levels of a single section of fine grained Tortonian marine clays near the village of Skouloudhiana in western Crete. The VGP paths corresponding to these two transitions, which are separated in time by about 1 Myr, are almost identical. The two VGP paths are largely constrained in longitude along a mean great circle about 80° west of the site, and thus cannot be classified as near- or far-sided. The same VGP path has also been observed for an R→N transition recorded in a nearby section at Potamida (Valet and Laj 1981). However the following N→R transition recorded at Potamida has a quite different VGP path, about 135° away from the R→N transition. This is not in accord with the standing field model which the coincidence of the Skouloudhiana paths would favour. The possible relation between excursions and transition fields is discussed in §4.6.

Rather than simulating the VGP path, Williams and Fuller (1981b) analysed inclination records associated with a given reversal. The basic assumption of their model of the transition field is that the dipole field decays in a nearly exponential fashion with part of the energy lost appearing in the first three terms (g_2^0, g_3^0, g_4^0) of the zonal non-dipole field. The decay of the dipole field energy was modelled on the intensity changes observed in the record of the Tatoosh intrusion (Dodson *et al* 1978). Williams and Fuller then calculated

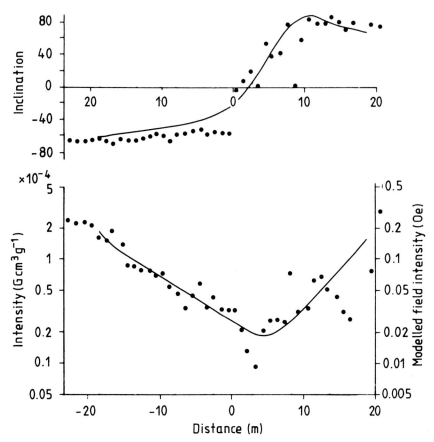

Figure 3.20 Comparison of inclination and intensity records from the Tatoosh intrusion, Washington (Dodson *et al* 1978) with the model for the Brunhes–Matuyama boundary. The site location is 46.8°N, 121.8°W. This record is a R→N reversal dated at 17 Myr recorded in a granodioritic intrusion. The modelled intensity is shown in Oe and the Tatoosh intensity data are for a mean of samples demagnetised to 100 Oe (after Williams and Fuller 1981b).

synthetic inclination and intensity records corresponding to the fields which would be observed if the transition field had the same harmonic content as the model. The last reversal (the Brunhes–Matuyama) was well simulated at most sites if the dipole energy is redistributed in g_2^0, g_3^0, g_4^0 in the ratio 2:3:5. Figure 3.20 shows the intensity and directional changes predicted by the model and the observed changes in the Tatoosh record. Both the model and observed data give a change in direction with a shorter timescale than the intensity change. However both intensity and inclination changes and estimates of the time to complete a reversal are strongly dependent upon latitude. Observations in the northern hemisphere indicate that intensity changes take much longer than directional ones (the Tatoosh intrusion is about 47°N), but at 45°S there is a gradual change in inclination that continues for almost the total duration of the event. This may explain why Opdyke *et al* (1973) found a comparable time for both the intensity and direction changes for the lower Jaramillo transition recorded in a southern hemisphere oceanic core (36°S).

As discussed in §3.3, the Agno batholith R→N reversal is atypical of R→N northern hemisphere reversals in that it is far-sided showing that the transition field geometry is not the same for all reversals. In a later paper, Williams and Fuller (1982) showed that if the transition field geometry is controlled by a single low-order zonal harmonic (either a quadrupole or an octupole field), the far-sided behaviour of the Agno record indicates that, for a given sense of reversal, the initiation hemisphere, or latitudinal band, cannot be constant. A quadrupole field initiated in the northern hemisphere would produce a far-sided path. However this is very different from the intermediate fields proposed by Hoffman (1981) for the last reversal. On the other hand, for a dominant octupole field to cause far-sidedness, initiation would have to occur simultaneously in both polar regions, which is rather unlikely. The same problem arises (initiation at both poles) if one modifies the harmonic model of Williams and Fuller (1981b) to produce far-sidedness by adjusting the distribution of energy among the various harmonics.

Gubbins and Roberts (1983) have used the frozen flux approximation (i.e. treating the core as a perfect conductor) to investigate transition paths. The validity of such an approximation is questionable however when applied to long timespans. It is a very powerful constraint because the topology of the null flux curves must be preserved. If the unsigned flux integral is held constant, it constrains the Gauss coefficients in a similar fashion to the energy integral of Williams and Fuller (1981b). However not all of the allowed configurations will satisfy the frozen flux requirement because some will lead to new null flux curves, and others to merging of null flux curves, neither of which are allowed. Features of transition fields are a large drop in intensity and axial symmetry.

Gubbins and Roberts (1983) showed that the intensity drop is possible with frozen flux, but raise the question as to whether it is possible to progress from a dipolar configuration, through an entirely axisymmetric transition field, to a reversed dipole. Although the unsigned flux integral could be preserved by a variety of low-order spherical harmonic series, giving transition paths like those in Williams and Fuller (1981b), none of them preserve the initial, single null flux curve. For axisymmetric fields the null flux curves are lines of latitude, and it is clearly impossible to change from, say, positive flux in the northern hemisphere and negative in the southern hemisphere, to the reverse situation, while keeping just one null flux curve on a line of latitude. Gubbins and Roberts give simple examples to show that it is possible to reverse the dipole (change the sign of g_1^0) if one begins with a field that is not a pure dipole. If non-

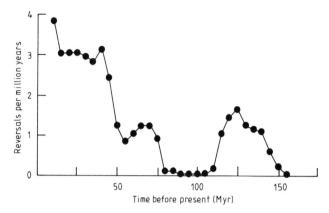

Figure 3.21 Variation in the average frequency of reversals as seen through a sliding window 10 Myr long. From 76 Myr to the present the rates are from Heirtzler *et al* (1968) and prior to 76 Myr from Larson and Pitman (1972) (after Cox 1975).

axisymmetric fields are allowed, it is trivial to achieve a reversal—a dipole can reverse simply by rotation, a process that is quite unrestricted by the flux integrals. Solutions of the α dynamo equations do not conserve flux as the electrical conductivity becomes large. This feature arises because the α term contains averaged effects involving diffusion of small-scale or asymmetric fields. Thus α dynamo models of transitions cannot be reconciled with frozen flux models.

3.5 Changes in the mean frequency of reversals

Figure 3.21 (after Cox 1975) shows the average frequency of reversals over the past 150 Myr based on the timescale of Heirtzler *et al* (1968) as seen through a sliding window 10 Myr long. Additional newly discovered short intervals could increase the average frequency by about 50% but should not significantly change the general pattern. Figure 3.21 shows that the mean frequency of reversals shows no statistically significant changes from about 75 to 45 Myr BP and then increases (rather abruptly) by a factor of more than two. It has then remained approximately constant until the present. The geomagnetic dynamo appears to have been stationary for the 30 Myr prior to this 45 Myr 'discontinuity' and to have been stationary since then. Thus 45 Myr seems to mark a boundary between two intervals during which the statistical properties of the dynamo were distinctly different.

An even more striking discontinuity appears at 85 Myr: for the preceding 22 Myr the frequency of reversals was zero or nearly so, the polarity of the field

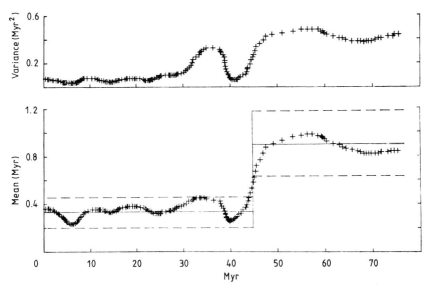

Figure 3.22   Mean and variance of interval lengths as a function of time for the Cenozoic reversal timescale of Heirtzler *et al* (1968). Units are millions of years (Myr). Plusses show the averages obtained with a Gaussian weighting function with a 2σ width of 10 intervals. No intervals were rejected in the analysis. Average means on either side of the discontinuity at 45 Myr are shown by horizontal lines. Broken lines define 95 % confidence regions about these averages (after Phillips *et al* 1975).

remaining normal from about 107 to 85 Myr BP. During the past 10^8 yr the frequency of reversals has thus increased from zero to the present frequency of nearly 5 per Myr in two steps, one at 85 Myr and one at 45 Myr. There have been other similar increases and decreases during the Earth's history (McElhinny 1971), changes in reversal frequency typically occurring at time intervals of 5×10^7 yr. Phillips *et al* (1975) have estimated the mean and variance of interval lengths as a function of time for the Cenozoic reversal timescale (0–76 Myr BP) of Heirtzler *et al* (1968) (figure 3.22). The figure clearly shows a discontinuity around 45 Myr. However, both curves are consistent with stationary behaviour between 0 and 45 Myr and between 45 and 76 Myr. Figure 3.23 shows the results for a similar analysis for the Keathley reversal sequence (116–153 Myr BP) of the Mesozoic reversal timescale of Larson (1974). It can be seen that this sequence is fundamentally different from the two earlier sequences. There appears to have been continuous non-stationary behaviour of the geodynamo during the Mesozoic.

The characteristic time associated with individual reversals (2×10^4 yr) differs by several orders of magnitude from that associated with changes in reversal frequency (5×10^7 yr). It seems probable therefore that the two phenomena have different causes. Individual reversals are more likely to have

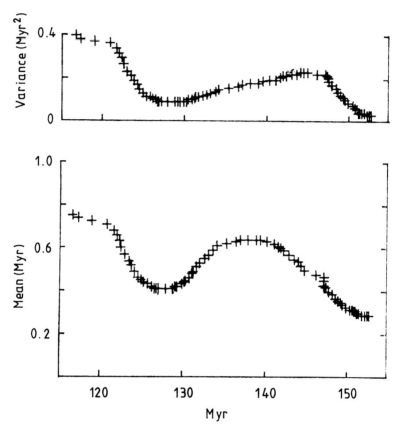

Figure 3.23   Mean and variance of interval lengths as a function of time for the Keathley reversal sequence of Larson's (1974) Mesozoic reversal timescale (after Phillips *et al* 1975).

their origin in the fluid motions and electric currents in the Earth's core—their characteristic times are those intrinsic to the geodynamo. Changes in reversal frequency, on the other hand, are probably due to changes in the rate at which energy becomes available to generate turbulence in the core or to a change in conditions at the MCB. Changes in the rate of supply of energy could result from processes such as the slow decay of radioactive heat sources in the core or to changes in the amount of latent heat of fusion supplied by growth of the solid inner core. The characteristic times of such processes are difficult to estimate but probably would be comparable with the age of the Earth. A time period of the order of 5×10^7 yr, on the other hand, is of the same order as that associated with geologic events—the formation of new ocean basins and mountain ranges. It is also a reasonable time to associate with convection in the mantle. The basic question is whether convection takes place through the mantle and, if so, whether it could affect the characteristics of the geodynamo.

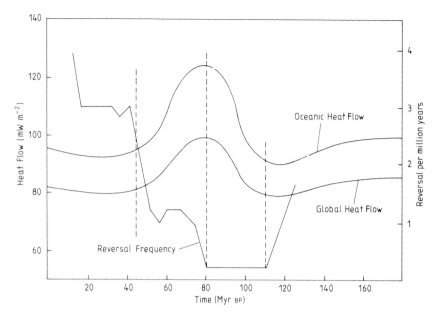

Figure 3.24 Variability of mean oceanic and global heat flow (after Sprague and Pollack 1980) and variations in the average frequency of reversals of the magnetic field (after Cox 1975) (after Jacobs 1981).

It is possible that deep mantle convection could cause hot and cold spots at the MCB (Cox and Doell 1964) and/or bumps on the MCB (Hide 1967); e.g. a cold descending column could produce both a cold spot and a bump due to the greater density of the material in the column. Jacobs (1981) showed that major changes in the frequency of reversals over the last 150 Myr occurred when there were changes in the heat flow from the Earth (see figure 3.24). Using magnetic data from Cox (1975) and heat flow data from Sprague and Pollack[†] (1980), he showed that the onset of the Cretaceous normal polarity interval about 105 Myr ago, during which there were almost no reversals, coincides with increased heat flow. This period of normal polarity lasted until about 85 Myr ago which is just about the maximum in the heat flow. Further, the sharp increase in the frequency of reversals about 45 Myr ago coincides with the next minimum in heat flow.

There is some evidence for a correlation between major tectonic changes and changes in reversal frequency. Figure 3.25 shows the magnetic reversal chronology for the last 80 Myr laid out alongside the Hawaii–Emperor chain. It can be seen that there is a major change in direction of the Pacific/mantle motion which occurred around 42–44 Myr ago, producing the Hawaii–

† Sprague and Pollack used the empirical relationships between terrestrial heat flux and the age of oceanic and continental crust and estimated the percentage of global area represented by crust of different ages.

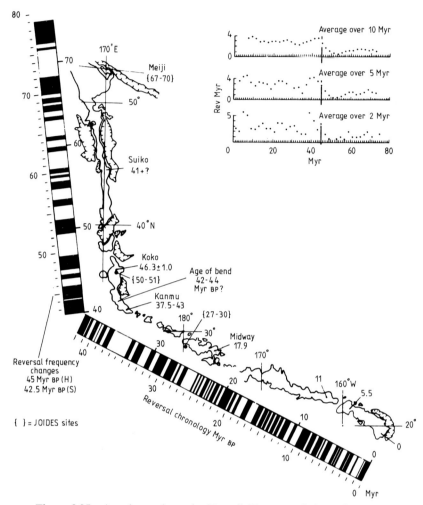

Figure 3.25 Age dates along the Hawaii–Emperor chain and reversal chronology, with two age scales along the lower and upper edges. Smoothed reversal frequency (Heirtzler *et al* 1968) shown in upper right with solid vertical line marking time of major frequency change about 45 Myr ago, which is also about the age of the Hawaii–Emperor Bend. Suggested revision of Heirtzler *et al* (1968) timescale moves the frequency change from 45 Myr to 42.5 Myr ago (after Vogt 1975).

Emperor Bend. This date coincides within the limits of probable error with the major change in reversal frequency around 45 Myr. Again 42–45 Myr marks the largest change in direction of plate motion between Greenland and Europe (Vogt and Avery 1974). Vogt (1975) has also shown that major tectonic changes occurred about 77 and 115 Myr ago, coinciding, within the dating uncertainty, with changes in reversal frequency.

Hide (1967) had earlier suggested that changes in the radial velocity of the fluid motions in the Earth's core might in some cases be impressed from outside. Horizontal temperature variations of only a few degrees and topological features, 'bumps' only a few kilometres high at the MCB might affect core motions, perhaps causing reversals. Gradual changes in the radius of the core and in the strength of the mechanism that drives core motions might also produce occasional reversals. Hide thus suggested that there are two types of reversals: 'forced' reversals due to changes impressed from outside the core, and 'free' reversals that would arise even in the absence of impressed changes. Each type of reversal would be characterised by its own timescale. Major geological events are associated with large-scale motions in the mantle. If these motions penetrate to a sufficient depth to produce horizontal variations in the physical conditions that prevail at the MCB, then 'forced' reversals should be strongly correlated with other worldwide geological phenomena. 'Free' reversals, however, should show no such correlations, being determined by random processes in the fluid core. These questions are discussed further in §5.4.

Lowrie (1982) has estimated the number of reversals per Myr in the Cenozoic and late Cretaceous averaged over intervals of 2, 5 and 10 Myr duration based on the timescale of Lowrie and Alvarez (1981). The 2 Myr averages show large, apparently irregular, fluctuations in reversal frequency; averaging over 5 Myr and especially over 10 Myr intervals indicates that these fluctuations are superposed on an almost linear trend (see figure 3.26). This suggests that the average reversal frequency of the geomagnetic field has been steadily increasing since the late Cretaceous. Lowrie and Kent (1983) repeated the calculations using different polarity timescales and an 8 Myr sliding window and arrived at similar conclusions. The difference between their interpretation of reversal frequency and that of Cox (1975) is discussed further in §5.5.

Mazaud *et al* (1983) have analysed the structure of the reversal frequency curve using the timescale of Lowrie and Alvarez (1981) and that of LaBrecque *et al* (1977). The frequency of reversals was studied using a moving rectangular window, with window widths ranging from 2 to 10 Myr. The reversal frequency curves were then analysed in terms of a monotonically increasing component and an oscillating component. The monotonic component was modelled by a least-squares fit to a Lorentzian function. To analyse the oscillating part they subtracted the corresponding Lorentzian from the reversal frequency curve and performed a time autocorrelation on the remainder and found a periodicity of about 15 Myr. They then fitted the reversal frequency curve simultaneously with a Lorentzian function and a sine function. For the 4 Myr window and the LaBrecque *et al* timescale, this gave a 31 Myr half width for the Lorentzian and a period of 15.1 Myr for the sine

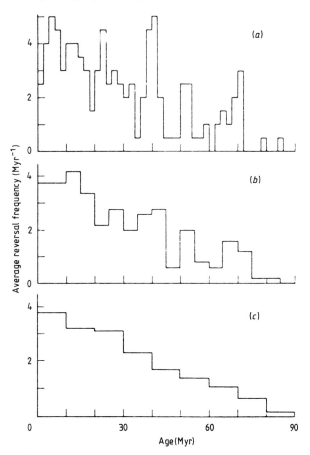

Figure 3.26 The frequency of geomagnetic polarity reversals since the late Cretaceous, averaged over intervals of (a) 2, (b) 5 and (c) 10 Myr respectively (after Lowrie 1982).

function (see figure 3.27). Similar results were found for windows ranging from 2 to 8 Myr and for the timescale of Lowrie and Alvarez.

McFadden (1984) does not believe that the fluctuations seen by Lowrie and Kent (1983) or those of Mazaud *et al* (1983) are real, but that the frequency of reversals has increased smoothly and linearly with time. He maintains that the superimposed fluctuations seen by Lowrie and Kent are due to their choice of smoothing filter (8 Myr) and those of Mazaud *et al* are an artifact of their analysis, the filter generating the fluctuations. It is just by chance that the period of the fluctuations seen by these two sets of authors are approximately the same.

The shapes of marine magnetic anomalies do not always agree with the shapes produced by the conventional two-dimensional block model of the

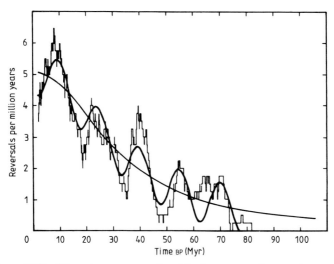

Figure 3.27   Direct least-squares fit of the frequency of reversals (on the LaBrecque *et al* (1977) timescale) using a Lorentzian function (after Mazaud *et al* 1983).

magnetic source. This was first pointed out by Weissel and Hayes (1972) who noticed that the remanent inclination inferred from the shapes of anomalies 5 to 20 on the north flank of the Southeast Indian Ocean Ridge was roughly 20° shallower than that inferred from the shapes of the same anomalies on the south flank. Cande (1976) later found a similar skewness discrepancy in the shapes of anomalies 27 to 32 in the North and South Pacific and referred to it as 'anomalous skewness'. More recently, Cande and Kristofferson (1977) discovered that the shapes of anomalies 33 and 34 in the North Atlantic are anomalously skewed by 40° on both sides of the Mid-Atlantic Ridge. This skewness discrepancy is larger by a factor of 2 or 3 than those previously observed. Cande (1978) showed that this large skewness of anomalies 33 and 34 is characteristic throughout the North and South Atlantic—west of the ridge crest they are skewed to the left, east of the ridge the conjugate anomalies are skewed to the right.

He initially (1976) suggested two ways of explaining skewness discrepancies. One is to modify the structure of the source layer assumed in the standard model, either by the physical rotation of crustal blocks containing the magnetic source, or by the addition of a deeper less strongly magnetised layer with broad sloping transition zones. The other is to modify the behaviour of the palaeomagnetic field assumed in the simple model, either by assuming that the average intensity of the palaeomagnetic field gradually decreases following a reversal or by assuming that there is an increase in the frequency or duration of undetected short polarity events towards the end of long periods of predominantly one polarity. Cande later (1978) showed that the second possibility—palaeomagnetic field behaviour—is the most likely. Lowrie and Alvarez (1977) have identified normal and reversed polarity zones corresponding to anomalies 34 through 29 in a sequence of upper Cretaceous limestones at Gubbio, Italy (see also §6.4). They found no evidence for previously undetected short polarity events within this entire sequence. This suggests that long periods of time of predominantly one polarity with very short events that increase in frequency did not occur at this time. Cande thus prefers the long-period intensity variation model.

One model for the generation of the Earth's magnetic field proposes that reversals occur as the result of a Poisson process (Cox 1969; see §5.5). An implication of this model is that the core has no memory longer than about 10^4 yr (Cox 1975). Since the average interval between reversals is of the order of 3×10^5 yr, the Poisson process model predicts that there should be no systematic trends in the field between reversals. Since the duration of the long reversed and long normal polarity intervals following anomaly 34 are approximately 3 and 5×10^6 yr respectively, Cande's (1978) explanation for the skewness of anomalies 33 and 34 reflects systematic trends in the field corresponding to core processes with a time constant of the order of 10^6 conflicting with the Poisson-process model.

References

Baag C G and Helsley C E 1974a Penecontemporaneous magnetisation of Moenkopi Formation *J. Geophys. Res.* **79** 3308

——1974b Remanent magnetisation of a 50 m core from the Moenkopi Formation, Western Colorado *Geophys. J.* **37** 245

Barbetti M F and McElhinny M W 1976 The Lake Mungo geomagnetic excursion *Phil. Trans. R. Soc.* A **281** 515

Bingham D K and Evans M E 1975 Precambrian geomagnetic field reversal *Nature* **253** 332

Blakely R J 1974 Geomagnetic reversals and crustal spreading rates during the Miocene *J. Geophys. Res.* **79** 2979

Blakely J R and Cox A 1972 Evidence for short geomagnetic polarity intervals in the early Cenozoic *J. Geophys. Res.* **77** 7065

Bogue S W and Coe R S 1982 Successive palaeomagnetic reversal records from Kauai *Nature* **295** 399

Bol'shakov A S and Solodovnikov G M 1980 Palaeomagnetic data on the intensity of the magnetic field of the Earth *Izv. Earth Phys.* **16** 602

Burakov K S, Gurary G Z, Khramov A N, Petrova G N, Rassanova G V and Rodionov V P 1976 Some peculiarities of the virtual pole positions during reversals *J. Geomag. Geoelec.* **28** 295

Cande S C 1976 A palaeomagnetic pole from Late Cretaceous marine magnetic anomalies in the Pacific *Geophys. J.* **41** 547

——1978 Anomalous behaviour of the palaeomagnetic field inferred from the skewness of anomalies 33 and 34 *Earth Planet. Sci. Lett.* **40** 275

Cande S C and Kristofferson Y 1977 Late Cretaceous magnetic anomalies in the North Atlantic *Earth Planet. Sci. Lett.* **35** 215

Clement B M, Kent D V and Opdyke N D 1982 Brunhes–Matuyama polarity transition in three deep-sea sediment cores *Phil. Trans. R. Soc.* A **306** 113

Cox A 1969 Geomagnetic reversals *Science* **163** 237

——1975 The frequency of geomagnetic reversals and the symmetry of the non-dipole field *Rev. Geophys. Space Phys.* **13** 35

Cox A and Doell R R 1964 Long period variations of the geomagnetic field *Bull. Seism. Soc. Am.* **54** 2243

Cox A, Doell R R and Dalrymple G R 1964 Reversals of the Earth's magnetic field *Science* **144** 1537

Creer K M 1972 The behaviour of the palaeogeomagnetic field during reversals *Trans. Am. Geophys. Union* **53** 614

Creer K M, Gross D L and Lineback J A 1976 Origin of regional geomagnetic variations recorded by Wisconsinan and Holocene sediments from Lake Michigan, USA and Lake Windermere, England *Geol. Soc. Am. Bull.* **87** 531

Creer K M and Ispir Y 1970 An interpretation of the behaviour of the geomagnetic field during polarity transitions *Phys. Earth Planet. Int.* **2** 283

Dagley P and Lawley E 1974 Palaeomagnetic evidence for the transitional behaviour of the geomagnetic field *Geophys. J.* **36** 577

Denham C R, Anderson R F and Bacon M P 1977 Palaeomagnetic and radiochemical age estimates for late Brunhes polarity episodes *Earth Planet. Sci. Lett.* **35** 384

Dodson R, Dunn J R, Fuller M, Williams I, Ito H, Schmidt V A and Wu Yu M 1978 Palaeomagnetic record of a late Tertiary field reversal *Geophys. J.* **53** 373

Dodson R, Fuller M and Kean W E 1977 Palaeomagnetic records of secular variation from Lake Michigan sediment cores *Earth Planet. Sci. Lett.* **34** 387

Doell R R and Cox A 1971 Pacific geomagnetic secular variation *Science* **171** 248

——1972 The Pacific geomagnetic secular variation anomaly and the question of lateral uniformity in the lower mantle *The Nature of the Solid Earth* ed E C Robertson (New York: McGraw-Hill)

Doell R and Dalrymple G B 1966 Geomagnetic polarity epochs; a new polarity event and the age of the Brunhes/Matuyama boundary *Science* **152** 1060

Dunn J R, Fuller M, Ito H and Schmidt V A 1971 Palaeomagnetic study of a reversal of the Earth's magnetic field *Science* **172** 840

Einarsson T and Sigurgeirsson T 1955 Rock magnetism in Iceland *Nature* **175** 892

Freed W K 1977 The vertical geomagnetic polepath during the Brunhes/Matuyama polarity change when viewed from equatorial latitudes *Trans. Am. Geophys. Union* **58** 380

Fuller M, Williams I and Hoffman K A 1979 Palaeomagnetic records of geomagnetic field reversals and the morphology of the transitional fields *Rev. Geophys. Space Phys.* **17** 179

Gubbins D and Roberts N 1983 Use of the frozen flux approximation in the interpretation of archaeomagnetic and palaeomagnetic data *Geophys. J.* **73** 675

Gurarii G Z 1981 The Matuyama–Jaramillo geomagnetic inversion in Western Turkmenia *Izv. Earth Phys.* **17** 212

Hammond S R, Seyb S M and Theyer F 1979 Geomagnetic polarity transitions in two oriented sediment cores from the northwest Pacific *Earth Planet. Sci. Lett.* **44** 167

Harrison C G A, McDougall I and Watkins N D 1979 A geomagnetic field reversal timescale back to 130 million years before present *Earth Planet. Sci. Lett.* **42** 143

Heirtzler J R, Dickson G O, Herron E M, Pitman W C III and Lepichon X 1968 Marine magnetic anomalies, geomagnetic field reversals and motions of the ocean floor and continents *J. Geophys. Res.* **73** 2119

Herrero-Bervera E and Helsley C E 1983 Palaeomagnetism of a polarity transition in the Lower (?) Triassic Chugwater Formation, Wyoming *J. Geophys. Res.* **88** 3506

Hide R 1967 Motion of the Earth's core and mantle, and variations of the main geomagnetic field *Science* **157** 55

——1981 Self-exciting dynamos and geomagnetic polarity changes *Nature* **293** 728

Hillhouse J and Cox A 1976 Brunhes–Matuyama polarity transition *Earth Planet. Sci. Lett.* **29** 51

Hillhouse J W, Cox A, Denham C R, Blakely R J and Butler R F 1972 Geomagnetic polarity transitions *Trans. Am. Geophys. Union* **53** 971

Hoffman K A 1977 Polarity transition records and the geomagnetic dynamo *Science* **196** 1329

——1979 Behaviour of the geodynamo during reversal: a phenomenological model *Earth Planet. Sci. Lett.* **44** 7

——1981 Quantitative description of the geomagnetic field during the Matuyama–Brunhes polarity transition *Phys. Earth Planet. Int.* **24** 229

——1982 The testing of geomagnetic reversal models: recent developments *Phil. Trans. R. Soc.* A **306** 147

Hoffman K A and Fuller M 1978 Transitional field configuration and geomagnetic reversal *Nature* **273** 715

Irving E and Pullaiah G 1976 Reversals of the geomagnetic field, magnetostratigraphy, and relative magnitude of palaeosecular variation in the Phanerozoic *Earth Sci. Rev.* **12** 35

Ito H 1970 Polarity transitions of the geomagnetic field deduced from the natural remanent magnetisation of Tertiary and Quaternary rocks in Southwest Japan *J. Geomag. Geoelec.* **22** 273

Jacobs J A 1981 Heat flow and reversals of the Earth's magnetic field *J. Geomag. Geolec.* **33** 527

Kaporovich I G, Makarova Z V, Petrova G N and Rybak R S 1966 The transitional stage of the geomagnetic field in the Pliocene on the territory of Turkmenia and Azerbaidzhan *Izv. Earth Phys.* **1** 59

Kawai N and Nakajima T 1975 Vanished geomagnetism *Proc. Jap. Acad.* **51** 640

Kawai N, Nakajima T, Hirooka K and Kobayashi K 1973 The oscillation of field in the Matuyama geomagnetic epoch and the fine structure of the geomagnetic transition *Rock magnetism and Palaeogeophysics* ed M Kono **1** 53

Kawai N, Otofuji Y and Kobayashi K 1976 Palaeomagnetic study of deep-sea sediments using thin sections *J. Geomag. Geoelec.* **28** 395

Kawai N, Sato T, Sueishi T and Kobayashi K 1977 Palaeomagnetic study of deep-sea sediments from the Melanesia Basin *J. Geomag. Geoelec.* **29** 211

Kristjansson L and McDougall I 1982 Some aspects of the late Tertiary geomagnetic field in Iceland *Geophys. J.* **68** 273

LaBrecque J L, Kent D V and Cande S C 1977 Revised magnetic polarity timescale for Late Cretaceous and Cenozoic time *Geology* **5** 330

Larson R L 1974 An updated timescale of magnetic reversals for the late Mesozoic *Trans. Am. Geophys. Union* **55** 236

Larson R L and Pitman W C 1972 World-wide Correlation of Mesozoic magnetic anomalies and its implications *Bull. Geol. Soc. Am.* **83** 3645

Lawley E A 1970 The intensity of the geomagnetic field in Iceland during Neogene polarity transitions and systematic deviations *Earth Planet. Sci. Lett.* **10** 145

Liddicoat J C 1981 Gauss/Matuyama polarity transition *Trans. Am. Geophys. Union* **62** 263

——1982 Gauss/Matuyama polarity transition *Phil. Trans. R. Soc.* A **306** 121

Lowrie W 1982 A revised magnetic polarity timescale for the Cretaceous and Cainozoic *Phil. Trans. R. Soc.* A **306** 129

Lowrie M and Alvarez W 1977 Upper-Cretaceous–Palaeocene magnetic stratigraphy at Gubbio, Italy, III. Upper Cretaceous magnetic stratigraphy *Geol. Soc. Am. Bull.* **88** 374

——1981 One hundred million years of geomagnetic polarity history *Geology* **9** 392

Lowrie W and Kent D V 1983 Geomagnetic reversal frequency since the Late Cretaceous *Earth Planet. Sci. Lett.* **62** 305

McDougall I, Saemundsson K, Johannesson H, Watkins N D and Kristjansson L 1977 Extension of the geomagnetic polarity time scale to 65 Myr: K–Ar dating, geological and palaeomagnetic study of a 3500 m lava succession in Western Iceland *Geol. Soc. Am. Bull.* **88** 1

McElhinny M W 1971 Geomagnetic reversals during the Phanerozoic *Science* **172** 157

McFadden P L 1984 15-Myr periodicity in the frequency of Geomagnetic reversals since 100 Myr *Nature* submitted

Mankinen E A, Donelly-Nolan J M, Grommé C S and Hearn B C Jr 1981 *US Geol. Surv. Prof. Pap.* **1141** 67

Massey N W D 1979 Keweenawan palaeomagnetic reversals at Mamainse Point, Ontario: fault repetition or three reversals? *Can. J. Earth Sci.* **16** 373

Mazaud A, Laj C, de Sèze L and Verosub K L 1983 Evidence for periodicity in the reversal frequency during the last 100 Ma *Nature* **304** 328

Nevanlinna H and Pesonen L J 1983 Late Precambrian Keweenawan asymmetric polarities as analysed by axial offset dipole geomagnetic models *J. Geophys. Res.* **88** 645

Niitsuma N 1971 Detailed study of the sediments recording the Matuyama–Brunhes geomagnetic reversal *Tohoku Univ. Sci. Rep. 2nd Ser. (Geol.)* **43** 1

Opdyke N D 1969 The Jaramillo event as detailed in oceanic cores *The application of modern physics to the Earth and Planetary Interiors* ed S K Runcorn (New York: Wiley)

Opdyke N D, Glass B, Hays J D and Foster J 1966 Palaeomagnetic study of Antarctic deep-sea cores *Science* **154** 349

Opdyke N D, Kent D V and Lowrie W 1973 Details of magnetic polarity transitions recorded in a high deposition rate deep-sea core *Earth Planet. Sci. Lett.* **20** 315

Palmer H C 1970 Palaeomagnetism and correlation of some Middle Keweenawan rocks, Lake Superior *Can. J. Earth Sci.* **7** 1410

Pesonen L J and Halls H C 1983 Geomagnetic field intensity and reversal asymmetry in late Precambrian Keweenawan rocks *Geophys. J.* **73** 241

Petrova G N and Rassanova G V 1976 Features of reversals *Principal geomagnetic field and problems of palaeomagnetism* Part II (Moscow: Nauka)

Phillips J D, Blakely R J and Cox A 1975 Independence of geomagnetic polarity intervals *Geophys. J.* **43** 747

Prevot M 1977 Large intensity changes of the non-dipole field during a polarity transition *Phys. Earth Planet. Int.* **13** 342

Robertson W A 1973 Pole positions from the Mamainse Point lavas and their bearing on a Keweenawan pole and polarity sequence *Can. J. Earth Sci.* **10** 1541

Shaw J 1975 Strong geomagnetic fields during a single Icelandic polarity transition *Geophys. J.* **40** 345

——1977 Further evidence for a strong intermediate state of the palaeomagnetic field *Geophys. J.* **48** 263

Shaw J, Dagley P and Mussett A 1982 The magnitude of the palaeomagnetic field in Iceland between 2 and 6 Myr ago *Geophys. J.* **68** 211

Shaw J and Wilson R L 1977 The magnitude of the palaeomagnetic field during a polarity transition: a new technique and its application *Phys. Earth Planet. Int.* **13** 339

Skiles D D 1970 A method of inferring the direction of drift of the geomagnetic field from palaeomagnetic data *J. Geomag. Geoelec.* **22** 441

Sprague D and Pollack H N 1980 Heat flow in the Mesozoic and Cenozoic *Nature* **285** 393

Steinhauser P and Vincenz S A 1973 Equatorial palaeopoles and behaviour of the dipole field during polarity transitions *Earth Planet. Sci. Lett.* **19** 113

Vadkovskii V N, Gurarii G Z and Mamikon'van M R 1980 Analysis of the process of geomagnetic field reversal *Izv. Earth Phys.* **16** 506

Valet J P and Laj C 1981 Palaeomagnetic record of two successive Miocene geomagnetic reversals in Western Crete *Earth Planet. Sci. Lett.* **54** 53

Valet J P, Laj C and Langereis C G 1983 Two different R–N geomagnetic reversals with identical VGP paths recorded at the same site *Nature* **304** 330

Van Zijl J S V, Graham K W T and Hales A L 1962 The palaeomagnetism of the Stormberg lavas, II. The behaviour of the magnetic field during a reversal *Geophys. J.* **7** 169

Vine F J 1966 Spreading of the ocean floor: new evidence *Science* **154** 1405

Vitorello I and Van der Voo R 1977 Magnetic stratigraphy of Lake Michigan sediments obtained from cores of lacustrine clay *Quat. Res.* **7** 398

Vogt P R 1975 Changes in geomagnetic reversal frequency at times of tectonic change: evidence for coupling between core and upper mantle processes *Earth Planet. Sci. Lett.* **25** 313

Vogt P R and Avery O E 1974 Detailed magnetic surveys in the Northeast Atlantic and Labrador Sea *J. Geophys. Res.* **79** 363

Watkins N D 1967 Unstable components and palaeomagnetic evidence for a geomagnetic polarity transition *J. Geomag. Geoelec.* **19** 63

——1969 Non-dipole behaviour during an upper Miocene geomagnetic polarity transition in Oregon *Geophys. J.* **17** 121

Watkins N D and Haggerty S E 1968 Oxidation and magnetic polarity in single Icelandic lavas and dykes *Geophys. J.* **15** 305

Weissel J K and Hayes D E 1972 Magnetic anomalies in the Southeast Indian Ocean *Antarctic Oceanology, II. The Australian–New Zealand sector* ed D E Hayes (Am. Geophys. Union, Antarctic Res. Ser.)

Williams I S and Fuller M 1981a A far-sided R→N VGP path from a reversal recorded in the Agno batholith *Trans. Am. Geophys. Union* **62** 853

——1981b Zonal harmonic models of reversal transition fields *J. Geophys. Res.* **86** 11657

——1982 A Miocene polarity transition (R→N) from the Agno Batholith, Luzon *J. Geophys. Res.* **87** 9408

Wilson R L 1972 Palaeomagnetic difference between normal and reversed field sources, and the problem of farsided and right-handed pole positions *Geophys. J.* **28** 295

Wilson R L, Dagley P and McCormack A G 1972 Palaeomagnetic evidence about the source of the geomagnetic field *Geophys. J.* **28** 213

4 Geomagnetic excursions

4.1 Introduction

In addition to polarity changes, the Earth's magnetic field has often departed for brief periods from its usual near-axial configuration, without establishing, and perhaps not even instantaneously approaching, a reversed direction. This type of behaviour has been called a geomagnetic excursion. Geomagnetic excursions have been reported in lava flows of various ages in different parts of the world and from some deep-sea and lake sediments. Excursions are generally observed to commence with a sudden and often fairly smooth movement of the VGP towards equatorial latitudes. The VGP may then return almost immediately, or it may cross the equator and move through latitudes in the oppoite hemisphere before swinging back again to resume a near-axial position. Barbetti and McElhinny (1976) define the term excursion to describe a VGP movement of more than 40° from the geographic pole (following the suggestion of Wilson *et al* (1972) for intermediate pole positions) which terminates with a return of the Earth's field to its pre-existing polarity, without the dynamo being observed to establish itself in the opposite polarity. Defined in this way excursions are differentiated from the secular variation (when the VGP colatitude $\theta < 40°$) and from short polarity events—a term Barbetti and McElhinny apply only when the opposite polarity ($\theta < 40°$ or $\theta > 140°$) persists long enough for at least one oscillation in the strength of the main dipole field (about 10^4 yr). It is possible that excursions represent abortive reversals. An example of two very short episodes in the Matuyama epoch seen in a deep-sea core taken in the western equatorial Pacific is shown in figure 4.1. A further example of short period events seen in other deep-sea cores in the Southern Ocean is shown in figure 4.2. Whether these events can be classified as excursions is not clear.

The youngest excursion proposed is the Starno event (Noel and Tarling 1975) which they inferred from a 30° shallowing in the inclination of demagnetised post-glacial sediments from Blekinge in southern Sweden. The low inclinations were seen in two samples from one core and in one sample

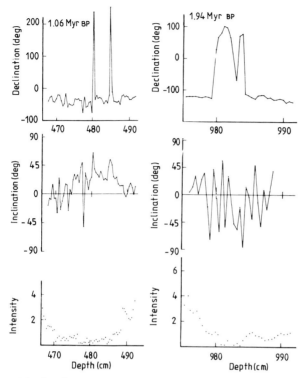

Figure 4.1 Declination, inclination and intensity of two short excursions as recorded in two deep-sea cores from the western equatorial Pacific in the Matuyama epoch (after Sueishi *et al* 1979).

from a second core at another site. Age correlation between the two sites is poor: about 1875 yr BP for the first site (Starno) and about 3700 yr BP for the second (Stilleryd). The mean of these two ages, 2800 yr BP, corresponds to that of a possible excursion suggested by Ransom (1973) on the basis of archaeomagnetic data from Italy and Greece. Although the archaeomagnetic data imply an inversion in the sign of the inclination, the lake sediment data of Noel and Tarling do not show this. Verosub and Banerjee (1977) have expressed grave doubts as to the reality of this excursion. It is extremely difficult to establish the reality of an excursion in every case — this point will be discussed again later.

4.2 The Laschamp excursion

One of the first geomagnetic excursions to be reported was that by Bonhommet and Babkine (1967) in the Laschamp and Olby lava flows from

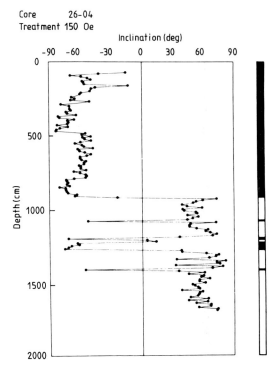

Core   26-04
Treatment 150 Oe

Figure 4.2 Inclination of short period events seen in a deep-sea core from the Southern Ocean. Polarity log at right: black is normal, clear is reversed (after Watkins 1968).

La Chaîne des Puys (Massif Central, France) and since called the Laschamp excursion. This still remains one of the best documented cases of a recent excursion. Unlike lake sediments, the Laschamp volcanics have a stable thermoremanent magnetisation which is not susceptible to structural degradation or detrital inclination error. The evidence for it in sediments is, however, conflicting. Noel and Tarling (1975) claim to have seen it in recent sediments although Denham and Cox (1971) and Thompson and Berglund (1976) have been unable to establish a correlation between it and sedimentary core samples. One of the main difficulties with such a correlation is the uncertainty of the age of the Laschamp and Olby lava flows which record the excursion. The first determination was by Bonhommet and Zahringer (1969) who gave an age between 8000 and 20 000 yr BP. Their lower limit was set by ^{14}C dating of baked trees found within a trachyte unit which overlies the Puy de Laschamp scoriae. As the scoriae and a nearby andesite flow both show the reverse magnetisation, the ^{14}C date is a minimum age for the event. The upper limit was found by whole-rock K–Ar dating of Laschamp and Olby flow samples. Because of the small quantities of radiogenic [40]Ar measured, Bonhommet and Zahringer gave only upper limits for the age, with a final value of 20 000 yr BP. Subsequently Hall and York (1978) using conventional whole-rock K–Ar analyses, obtained a weighted average age of $45\,400 \pm 2500$ yr BP. They attributed these consistently older ages to the difference in correction for atmospheric (or initial) argon contamination. With such young samples and such low $^{40}Ar/^{36}Ar$ ratios, the age is critically dependent on the assumed $^{40}Ar/^{36}Ar$ ratio. Hall and York also used the $^{40}Ar/^{39}Ar$ dating method with step heating and obtained an age of $47\,400 \pm 1900$ yr BP in excellent agreement with their results using conventional analyses. Condomines (1978) used the $^{230}Th/^{238}U$ radioactive disequilibrium method to date one of the reversely magnetised Olby flows, and obtained an age of $39\,000 \pm 6000$ yr BP, which is again considerably older than that originally given by Bonhommet and Zahringer.

The reversely magnetised Laschamp and Olby flows have also been dated by thermoluminescence (TL). An earlier attempt (Wintle 1973) was frustrated by anomalous fading of the TL signal. Such anomalous fading seems to be exhibited by most of the TL-sensitive minerals in lavas and Wintle (1977) suggested that it is probably due to wave mechanical leakage of electrons out of traps to nearby defects. Huxtable *et al* (1978) avoided this difficulty by using TL to date the clay and sediment baked by the lava flows. They obtained an age of $25\,800 \pm 1700$ yr BP in fair agreement with a TL age of $33\,000 \pm 4000$ yr BP obtained by Valladas *et al* (1977) based on quartz extracted from a granitic inclusion. Guérin and Valladas (1980) have also obtained ages for the Laschamp and Olby lava flows using TL techniques on plagioclase feldspars. They obtained ages of $32\,500 \pm 3100$ yr BP for Laschamp and $37\,300 \pm 3500$ yr BP for Olby. More recently, Guérin *et al* (1984) have dated these lava flows using high-temperature (800–1000 K) TL properties of plagioclase and K–Ar techniques. Their TL ages for the Laschamp and Olby events are $32\,500 \pm 3100$ and $37\,300 \pm 3500$ yr BP respectively and their K–Ar ages $38\,000 \pm 8000$ and $42\,000 \pm 9000$ yr BP.

There have been still more attempts to determine the age of the Laschamp and Olby reversely magnetised lava flows. Using whole-rock samples Gillot *et al* (1979) obtained a K–Ar age of $43\,000 \pm 5000$ yr BP for the Laschamp flow and $50\,000 \pm 7500$ yr BP for the Olby flow in excellent agreement with the ages obtained by Hall and York (1978). They also estimated the ages by TL. For the Laschamp flow they used quartz from the granitic inclusion and obtained an age of $35\,000 \pm 3000$ yr BP; the age of the Olby flow was measured from five quartz pebbles from the baked palaeosoil and gave an age of $38\,000 \pm 6000$ yr BP. They also obtained TL ages of plagioclases obtained from ground rock samples, and found an age of $33\,500 \pm 5000$ yr BP for the Laschamp flow and $44\,100 \pm 6500$ yr BP for the Olby flow. Both these ages are greater than the other TL ages obtained by Huxtable *et al* (1978). Gillot *et al* (1979) also obtained ^{14}C ages for these flows—the most probable age of the Olby organic layer was found to be greater than 36 000 yr.

Barbetti and Flude (1979a) have made measurements of the geomagnetic field strength of sediments baked by the lava flows from La Chaîne des Puys. For Royat the dipole moment was only 30% of its present value. The TL age for the Royat flow obtained by Huxtable *et al* (1978) was $25\,800 \pm 1700$ yr BP. However, Hall *et al* (1979) obtained a $^{40}Ar/^{39}Ar$ age of about 40 000 yr which is not significantly different from that of the Laschamp and Olby flows.

This intense amount of work on the dating of these flows underlines one of the main difficulties in correlating reported excursions of the Earth's magnetic field—namely that of determining reliable ages. In any theory of magnetic excursions it is essential to know whether they are worldwide events or are observed over only a small part of the globe. When the Laschamp excursion was originally dated as lying between 8000 and 20 000 yr BP (Bonhommet and Zahringer 1969), many attempts were made to correlate it with other excursions observed in other parts of the world, such as the Gothenburg (Morner and Lanser 1974), Gulf of Mexico (Clark and Kennett 1973), Lake Biwa (Nakajima *et al* 1973). Denham and Cox showed in 1971 that if the Olby–Laschamp excursion had affected the entire Earth's magnetic field, it could not have happened between 13 300 and 30 400 years ago or lasted less than 1700 years. Now that it is known that the event is much older, correlations have been sought with older reported excursions. Other magnetic field disturbances observed in the period 30 000–50 000 yr BP are the Lake Mungo excursion (Barbetti and McElhinny 1972, 1976) which lasted from 28 000 to at least 31 000 yr BP (see §4.3), the Lake Biwa excursion (Yaskawa *et al* 1973, Yaskawa 1974), dated by extrapolation at 49 000 yr BP, an excursion, the age of which is estimated at 40 000 yr BP in sediments from the Indian Ocean (Opdyke *et al* 1974) and possibly an event recently discovered in Iceland (the Maelifell event) (Peirce and Clark 1978) which has not yet been dated accurately. It is difficult to relate these latter 'events' to the Olby–Laschamp excursion. If, however, the Lake Mungo and Olby–Laschamp disturbances represent the same event, Peirce and Clark have shown that it must have lasted several thousand years $(11\,000 \pm 6000$ yr).

Heller (1980) and Heller and Petersen (1982) have carried out thermal laboratory experiments and observed complete or partial self-reversal of the NRM in many of the Olby samples and to a lesser extent in some of the Laschamp flows. After magnetic and optical examination of the samples, they suggested that the self-reversal mechanism is caused by a negative magnetostatic coupling between titanomagnetite phases of widely varying degrees of oxidation. Heller and Petersen also studied the magnetostratigraphy of a contemporaneous loess section at Steinheim in southern Germany and found no reversed polarities. They did observe some abrupt changes in declination of very short duration (less than 2000 yr) which they suggested are probably caused by mechanical disturbances (such as solifluction) rather than representing geomagnetic field variations. No positive evidence for the Laschamp event was detected in other loess profiles from Czechoslovakia and northern China.

This work indicates the need for thorough rock-magnetic investigations before ascribing every excursion to a reversal of the geomagnetic field.

4.3 The Lake Mungo excursion

Barbetti and McElhinny (1972, 1976) carried out archaeomagnetic studies of prehistoric aboriginal fireplaces along the ancient shore of Lake Mungo, a dried-out lake in south-eastern Australia. Directions of magnetisation preserved in oven stones and baked hearths showed that wide departures of up to $120°$ from the direction of the axial dipole field occurred about 30 000 years ago. The geomagnetic excursion recorded between at least $30\,780 \pm 520$ and $28\,140 \pm 370$ yr BP is associated with very high field strengths between 1 and 2×10^{-4} T (see §3.2). The field strength subsequently decreased to between 0.2 and 0.3×10^{-4} T after the excursion. This main excursion is referred to as the Lake Mungo excursion. There is some evidence that a second excursion, associated with low field strengths of $0.1–0.2 \times 10^{-4}$ T, occurred around 26 000 yr BP. These dates were obtained from ^{14}C measurements in charcoal collected from the fireplaces and are consistent with the stratigraphic evidence. Huxtable and Aitken (1977), using quartz grains extracted from the baked clay–sand matrix, obtained TL ages for the first excursion of $35\,000 \pm 4300$ yr BP. This age is not significantly different from the average of the ^{14}C dates. It is of interest that Oberg and Evans (1977) found no evidence of the Lake Mungo excursion in a seven metre sedimentary sequence in southern British Columbia.

Because most of the excursions proposed so far have been found in sediments and have not been convincingly documented over a sizable portion of the globe, several workers (Verosub 1975, Verosub and Banerjee 1977, Thompson and Berglund 1976) have suggested that some or most of them may reflect sedimentological rather than geomagnetic phenomena. The record of the Lake Mungo excursion, however, is contained in sedimentary material that was baked in prehistoric aboriginal fireplaces. Thus, sedimentological phenomena cannot be invoked to explain the anomalous directions found in this case because the NMR is not detrital or diagenetic in origin but rather is thermoremanent.

It is possible to simulate the record of a reversal in a number of ways, e.g. the dipole field terms can be reduced to zero and then allowed to grow with opposite polarity, while some attempt is made to simulate the behaviour of the non-dipole field. Larson *et al* (1971) carried out such a simulation using a number of subsidiary dipoles placed close to the MCB. These were allowed to drift past the site to simulate the westward drift of the secular variation. In this way they were able to generate a sequence of field directions at the site which were similar to those observed in the reverse-to-normal transition recorded in a sequence of Late Miocene volcanics dated at about 15 Myr in the Santa Rosa Range, north central Nevada. In a similar model of the Tatoosh reversal, Dodson *et al* (1977) found that to achieve a reasonable simulation it was necessary to maintain a fixed horizontal dipole term, which did not reverse at the same time as the axial dipole. Also, to obtain a path which is confined in longitude it was necessary to maintain some part of the non-dipole field with constant polarity throughout the reversal of the main dipole.

Harrison and Ramirez (1975) have modelled pseudo reversals (i.e. ones not involving the main dipole field) by assuming that they are caused by a dipole at the MCB, whose magnetic field is opposite to that of the main dipole field. They showed that the areal coverage of disturbed magnetic field around such a pseudo reversal can be quite small, so that observations made only a few thousand kilometres away would show no anomalous directions. If this model is correct, they further showed that pseudo reversals should be much less common at low latitudes than at high latitudes. Barbetti and McElhinny (1976) used such a model to explain the Lake Mungo excursion. A radial dipole (offset from the centre of the Earth by 0.5 Earth radii) would need a moment of only one-eighth that of the main geocentric dipole to produce at the point on the surface nearest to it a field greater than or equal to that of the main dipole, whereas on the far side of the Earth it would produce a field only 1/27 as large.

Coe (1977) considered various simple sources, all in the Earth's outer core, that could account for the Lake Mungo excursion — an eccentric radial dipole or current loop, an eccentric horizontal dipole, and a pair of eccentric radial current loops, of opposite sign. A horizontal eccentric source or current loop is the most efficient in terms of causing the least disturbance of the field elsewhere, but there is little basis for expecting one to exist since the modern non-dipole field is dominated by a few large features that are most simply modelled by radial sources. Coe showed, however, that single radial sources would produce extremely high fields over very large areas and significantly high anomalous fields over the entire globe. Thus they are improbable unless simultaneous worldwide effects are discovered. The best model seems to be a combination of two radial eccentric sources of opposite sign. In this case horizontal components add and the area dominated by their combined fields is much less than that from a single source. Coe showed that in their most efficient configuration such a pair would cause a maximum field of about 2.3×10^{-4} T while producing the required non-dipole field at Lake Mungo. However, significant effects would occur as far away as south-eastern Asia and India, southern Africa, and much of Antarctica and the south Pacific Basin.

4.4 The Mono Lake excursion

Denham and Cox (1971) carried out palaeomagnetic measurements on late Pleistocene sediments in Mono Lake, California, and found no evidence for a reversal between 30 400 and 13 300 yr BP with ages controlled by ^{14}C dating. If the Laschamp excursion occurred during this time, its duration can have been no longer than 1700 yr (the largest sampling gap in their data). However the sediments did record a well defined excursion in the direction of the field about 24 000 yr ago with a peak-to-peak amplitude of 25° and a period of about 600 yr. They attributed the anticlockwise looping motion of the magnetic field vector to a local eastward drift of the non-dipole field. Denham later (1974) investigated the records in more detail and showed that the major features of the loop could be explained by an inwardly directed radial dipole located at 0.5 Earth radii drifting eastwards along a path 15° south of Mono Lake. Its drift velocity was 0.10–0.19° yr^{-1} as compared to the present westward drift of about 0.2° yr^{-1} and its strength was 0.12–0.21 relative to the main dipole moment. He estimated the lifetime of the disturbance to be of the order of 850 yr.

Verosub (1977a) has pointed out that the presence of anomalous directions in only the inclination or declination but not in both is more likely to be caused by distortion in the palaeomagnetic recording process or by errors in the sampling procedure than by fluctuations in the geomagnetic field. A major problem is the difficulty in finding consistent records of geomagnetic excursions. In some cases a proposed magnetic excursion can be found at the same stratigraphic horizon but with greatly varied magnetic signature (Freed and Healy 1974): in other cases the magnetic feature may be entirely absent from nearby, contemporaneous sediments (Creer *et al* 1976a, b).

The most convincing evidence for the existence of the Mono Lake excursion is that it has been found at two sites seventeen kilometres apart within the same sedimentary environment (Denham 1974). On the other hand, sediments from Clear Lake, spanning the time interval 21 000–29 000 yr BP do not record any anomalous magnetic features (Verosub 1977a). Since Clear Lake is only 320 km from Mono Lake and since each sample from Clear Lake represents 26 years of sedimentation, the magnetic signature of the Mono Lake excursion should be recorded in detail in the Clear Lake samples. The absence of the Mono Lake feature from the palaeomagnetic record of Clear Lake makes it difficult to accept the Mono Lake record as a real geomagnetic excursion. If the reality of the Mono Lake geomagnetic excursion is not confirmed, it indicates that even the existence of magnetically consistent records from two widely separated sites within the same sedimentary environment is not sufficient to establish the existence of a geomagnetic excursion. On the other hand if the reality of the Mono Lake excursion is confirmed, its absence from the palaeomagnetic record of Clear Lake indicates that unfortunately geomagnetic excursions can have only limited reliability as magnetostratigraphic horizons for correlation between different sedimentary environments.

Liddicoat and Coe (1979) have since re-examined the Mono Lake excursion in more detail—extensive sampling revealed a new feature of the excursion. They found the previously known eastward swing in declination and steepening of inclination to be preceded by an even greater swing to westerly declination and shallow inclination. The duration of the entire excursion is about 1000 yr. Excellent agreement of palaeomagnetic directions between four sites shows that the excursion is a real expression of changes of the geomagnetic field. They modelled the source by a radial eccentric dipole at high northern latitudes pointing outward during the first part of the excursion, and near the equator and pointing inward during the latter part. The eccentric dipole was offset from the centre of the Earth by a distance of 0.28 Earth radii (approximately 1784 km). Movement of the source appears to be localised, displaying a complex pattern of eastward, westward, and even northward drift. The average moment of the hypothetical eccentric dipole during the excursion is comparable with the largest calculated for the 1945 field, and the maximum moment is almost twice as great. The ages of all the samples range from 30 700–23 350 yr BP based on ^{14}C dates and the assumption that the sedimentation rate was uniform. Their data are in general agreement with Denham and Cox's (1971) result of relatively quiet behaviour of field direction from about 30 700–25 000 yr BP followed by an excursion between about 25 000 and 24 000 yr BP. The answer to the Mono Lake, Clear Lake problem may be that part of the record is missing. Uniform sedimentation rates were assumed but some of the deposition may have been destroyed e.g. by bottom water currents. More probably however the answer lies in errors in ^{14}C dating (see e.g. Barton and Polach 1971 and §4.7).

Lajoie and Liddicoat (1980) later found large swings in declination (up to 60°) and inclination (up to 50°) in widely spaced samples from the post-excursion portion of the stratigraphic section record, confirming the broad swings reported previously. Five and a half clockwise loops span the set of smoothed VGP in the interval 29 000–13 000 yr BP. The loops have an average duration of 2900 yr and their movement is comparable with the westward drift of the non-dipole field. A new core has now been taken from Clear Lake which may shed further light on the controversy. The earlier core, which had partly dried up when the subsamples were taken, gave poor quality magnetic data.

Verosub *et al* (1980) have examined a sequence of deep lake clays exposed on the shores of Pyramid Lake, Nevada, 230 km from Mono Lake, covering the time interval 25 000–36 000 yr BP. These dates were obtained from ^{14}C measurements on two wood fragments and from one sample of disseminated organic material. The spread in ages is consistent with the assumption of a uniform and continuous sedimentation rate. The measured ranges of inclination and declination are 40° and 75° respectively. Thus the palaeomagnetic record from Pyramid Lake, like that from Clear Lake, contains no evidence for a geomagnetic excursion. Taken together it would seem that Northern California and Western Nevada were not affected by an excursion during the time period 21 000–36 000 yr BP. This is in direct conflict with the data from Mono Lake. Verosub *et al* (1980) have discussed this problem in detail. If the absence of the excursion at both Clear Lake and Pyramid Lake is attributed to non-deposition or erosion, such processes would have to have occurred simultaneously at each lake and simultaneously with the geomagnetic excursion. Such a coincidence is highly unlikely. The simplest explanation is, as already stated, that the original data of the Mono Lake excursion is in error. It should also be noted that studies of contemporaneous lake sediments by Oberg and Evans (1977) in southern British Columbia and by Doh and Steele (1981) in Fargher Lake, south-west Washington, failed to confirm the existence of an excursion. In a later paper, Doh and Steele (1983) confirmed that for the most part the sediments of Fargher Lake have a strong, stable remanent magnetisation showing no evidence of reversals. They concluded that none occurred during deposition except perhaps during two short gaps in their records totalling about 1700 yr for which they were unable to recover any sediments. If the Heussers' (1980) timescale is correct, no reversals occurred in south-western Washington during the time interval 14 100–30 300 yr BP unless it occurred during the intervals 15 800–16 800 yr BP or 21 100–21 800 yr BP.

Palmer *et al* (1979) have carried out palaeomagnetic and sedimentological studies at Lake Tahoe, California–Nevada and found fair agreement between the Mono Lake and Lake Tahoe records in both declination and inclination (see figure 4.3). The correlation is based on the similarity of longer wavelength

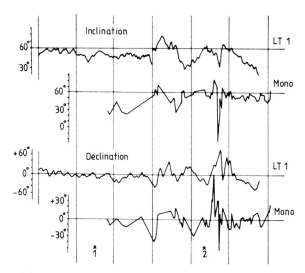

Figure 4.3 Proposed correlations of NRM inclination and declination data from Lake Tahoe and Mono Lake. Correlation is based upon the similarity of longer wavelength variations between points 1 (13 300 ± 500 yr) and 2 (23 300 ± 300 yr). On the basis of this correlation there is an alignment of a short-wavelength, high-amplitude feature in both the inclination and declination data from Lake Tahoe with the record of the Mono Lake geomagnetic excursion (after Palmer *et al* 1979).

variations. On the basis of this correlation the one short-wavelength high-amplitude feature on the Mono Lake record is aligned with a similar short wavelength feature common to all three cores from Lake Tahoe. There is however no definite age determination on the cores from Lake Tahoe and the Mono Lake excursion was fixed at 24 000 yr BP.

Liddicoat has looked at the Pyramid Lake records and suggested that perhaps there is a hint of an excursion near the middle of the record where it should be on the basis of the Mono Lake records. More recently, Liddicoat *et al* (1982) have identified the Mono Lake excursion in the Lake Lahontan Sehoo Formation in Carson Sink, Nevada, 200 km NE of Mono Basin, where it was originally documented in the Wilson Creek beds near the north-west shore of Mono Lake. The age of the Mono Lake excursion is placed at 28 000–26 000 yr BP, on the basis of interpolation from a linear regression of sixteen ^{14}C dates on nodular and platy algal tufa in the Wilson Creek beds. This age is consistent with a 24 480 ± 430 yr ^{14}C date on wood associated with the Wono tephra bed in Lake Lahontan lacustrine deposits near Pyramid Lake.

Turner *et al* (1982) sampled an eighteen metre sedimentary sequence at Bessette Creek, British Columbia (about 12° north of Mono Lake). Radiocarbon ages suggested that the sequence spans the interval 31 200–19 500 yr BP. No evidence for any large geomagnetic excursions were found, although a distinctive pattern of 'normal' secular variation was observed with declination and inclination swings of 45° and 25° peak-to-peak amplitude respectively. For the most part the secular variation consisted of low amplitude oscillations about the field vector of a geomagnetic axial dipole expected at the site latitude, but three relatively large perturbations occurred at approximately 4000 yr intervals.

4.5 The Gothenburg 'flip'

Morner *et al* (1971) and Morner and Lanser (1974) reported a very rapid 'excursion' in the magnetic field which they subsequently named the Gothenburg 'flip' (figure 4.4). They first recognised it in a core in the Botanical Gardens, Gothenburg, Sweden. The upper boundary was fixed very precisely at the boundary between the Fjaras Stadial and the Bolling Interstadial, dated at 12 350 yr BP. The change from reversed to normal polarity is very rapid, occurring within a few years! Furthermore, there is no major intensity change related to the polarity change. They speculated that there may have been a long period (1000–2000 yr) of unstable magnetism, with several flips between normal and reversed polarity, or a short period (about 100 yr) of reversed polarity at around the Fjaras Stadial (12 400–12 350 yr BP) which followed a millennium of irregular (but not fully reversed) magnetism. Morner (1977) later claimed that the excursion is also present at the same stratigraphic level in four other Swedish cores.

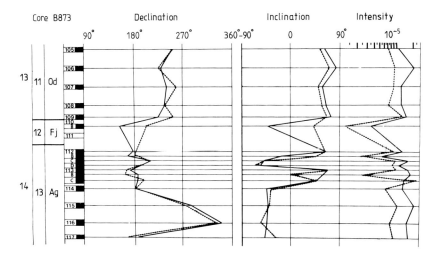

Figure 4.4 Palaeomagnetic measurements of the base of Core 873 taken in the Botanical Gardens, Gothenburg, Sweden. The 'flip' includes layers 12 and 13, that is, the top of the Agard Interstadial (Ag) and Fjaras Stadial (Fj). Full lines, natural remanent magnetisation; dotted lines, magnetisation after cleaning in 200 Oe. Layers 12 and 13 probably only represent about 100 yr (after Morner and Lanser 1974).

Thompson and Berglund (1976) believe that the intermediate palaeomagnetic directions observed by Morner and Lanser are due to slumping rather than a real change in the magnetic field, which they maintain was of normal polarity from 13 000–11 000 yr BP in Sweden. They base their view on an analysis of 408 subsamples taken from two cores in southern Sweden. Their data suggest that previously reported reversed palaeomagnetic directions are not reliable indicators of the ancient geomagnetic field, but have been distorted by mechanical sedimentation processes, slumping or weathering. They further suggested that the proliferation of unusual palaeomagnetic directions in Scandinavia around 12 000 yr BP is a reflection of changing climatic conditions. In many localities fluctuations of climate produced sediments of very variable mechanical properties, particularly at times of periglacial activity, which were poor recorders of the direction of the ambient magnetic field. Banerjee *et al* (1979) confirm this conclusion. An earlier report that the Gothenburg 'flip' had been observed in New Zealand has also since been refuted (Sukroo *et al* 1978).

Barbetti *et al* (1980) have successfully measured the ancient magnetic direction and field strength in baked stones from late Pleistocene hearths at Étiolles and Marsangy, France. The dates for two hearths at Étiolles have been estimated as 12 000 ± 220 yr BP, obtained by ^{14}C dating of a mammoth scapula excavated from a lower level in another part of the site. The only means of dating the stones at Marsangy is by archaeological seriation based mainly on the stone industry. Barbetti *et al* assumed an age of about $12\,000 \pm 500$ yr BP. They found the VGP to be very close to, and not significantly different from, the Earth's geographic pole, and the corresponding VDM to be only slightly less than the present-day dipole moment. These results strengthen the conclusions of Thompson and Berglund (1976), casting doubt on the reality of the Gothenburg excursion.

Morner and Lanser (1975) later analysed core A179-15—a high deposition rate core from the southern North Atlantic and found a sudden change in the declination at $12\,350$ yr BP which coincides exactly with the end of their Gothenburg flip. Opdyke (1976) does not accept their findings, believing that the signal seen in declination only is noise arising from the sampling procedures used in this old core. However, Morner (1976) does not accept Opdyke's criticism and is still convinced of the reality of the Gothenburg flip. Abrahamsen (1982) examined three sedimentary cores from Solberga, Brastad and Moltemyr, north of Gothenburg, Sweden. In the Solberga core magnetic directions between depths 17 and 12 m showed great scatter indicating a possible excursion of the magnetic field. However, no well documented excursions of the geomagnetic field in Holocene time are yet known, and Abrahamsen suggested that the directional scatter could most simply be explained as due to some kind of post-depositional disturbance such as sliding, slumping, bioturbation, or compaction.

There are several signs in the Solberga core around 18–19 m depth of a change in the environment from saline to more brackish water conditions. This may be related to the climatic amelioration of the Pleistocene–Holocene transition, with an increase in the meltwater flux and with the drainage of the Baltic Ice Lake. A change in the wet density is seen at 18.95 m and also a jump in the NRM intensity at that depth and again at 19.35 m. A change in declination is observed at 17.95 and 18.50 m but not in the cleaned records. The inclination shows a smooth variation below 17 m with a gradual increase between 20 and 17 m, indicating that at least up to this level, a post-depositional disturbance of the sediment is unlikely to have occurred.

The 'Gothenburg excursion' of the geomagnetic field, which Morner *et al* (1971) postulated to end at $12\,350 \pm 50$ yr BP is too early to appear in the Solberga core with an age at the bottom of $11\,200 \pm 400$ yr BP. The Brastad core, however, is likely to reach further back in time. Below 13.5 m the declinations and the inclinations are very scattered, and at the very bottom of the core, four specimens with low inclinations and ordinary declinations were found. During magnetic cleaning the scatter is not significantly altered in the bottom metre, indicating that viscous components do not influence the scattered directions. The directional scatter must therefore be due to orientational scatter in the sediment rather than a differential response to viscous overprints in the ambient field. Abrahamsen concluded that a geomagnetic excursion is probably seen at the bottom of the Brastad core.

During progressive AF demagnetisation, an interval of the Holocene part of both the Solberga and Moltemyr cores shows decreasing inclinations probably due to viscous magnetic overprinting, which may indicate either unusual magneto-sedimentological properties related to an increase in deposition rate, or a hitherto unrecognised post-glacial geomagnetic low inclination excursion, occurring shortly after 10 000 yr BP. The Gothenburg excursion, dated to end around 12 350 yr BP, is not seen in these records, although the Brastad and Moltemyr cores do probably reach further back.

4.6 Other excursions

The Laschamp, Lake Mungo, Mono Lake and Gothenburg excursions have received most attention in the literature, although many other possible excursions have been reported. Palaeomagnetic and micropalaeontological studies have been carried out by Clark and Kennett (1973) on 28 sedimentary cores from the Gulf of Mexico with sedimentation rates ranging from 9–20 cm/1000 yr. An excursion of the Earth's magnetic field was found in the upper parts of eight out of fifteen cores for which palaeomagnetic studies were conducted and was independently correlated with planktonic foraminiferal zones. The age of the excursion was determined by extrapolation of sedimentation rates from a palaeontological boundary and occurred between 12 500 and 17 000 yr BP. Clarke and Kennett attributed the excursion in several cores from the western Gulf of Mexico to the dominance of calcium carbonate. Almost all the cores from the eastern Gulf of Mexico that were used for palaeomagnetic measurements show the excursion. These cores were, except for the uppermost few centimetres, almost devoid of calcium carbonate. Freed and Healy (1974) have also examined deep-sea sediment cores from the Gulf of Mexico. They found two excursions, one at $17\,000 \pm 1500$ yr BP and the other at $32\,000 \pm 1500$ yr BP. The excursions were dated by microfaunal analysis.

Tanaka and Tachibana (1979, 1981) reported an excursion in a welded tuff at Shibutami, Japan. Geological and ^{14}C ages place the tuff at about 30 000 yr BP. They speculated about a possible correlation with other reported excursions at the time, but came to no conclusion. They also carried out intensity measurements—although possible errors were quite large, the intensity was small at the time the welded tuff was deposited (being only about 20% of the present day field). Tanaka and Tachibana (1981) noted that the younger Lake Mungo excursion (around 26 000 yr BP) is also associated with low magnetic field intensity.[†]

† This 'excursion' has since been discounted (Tanaka, personal communication). The age of the tuff by Tachibana has now been revised back to around 0.7 Myr (i.e. to about the time of the Brunhes–Matuyama boundary) by new geological evidence and K–Ar dating. This stresses the difficulty of interpreting correctly every possible reversal of the geomagnetic field.

Yaskawa *et al* (1973) carried out palaeomagnetic measurements in a 197.2 m core from Lake Biwa, Japan. Sedimentation rates in large lakes are usually high—in Lake Biwa as much as 500 m Myr^{-1}—nearly two orders of magnitude greater than those of most deep-sea sediments. The results of a preliminary survey indicate that there may have been five short reversals during the Brunhes normal epoch. The curve of inclination against depth correlates well with that obtained by Wollin *et al* (1971) in a deep-sea core from the North Pacific—this core had a timescale interpolated from the age of the Brunhes–Matuyama boundary (see figure 4.5).

From this correlation, Yaskawa *et al* suggested that the youngest of the three clear reversals is the same as the Blake 'event' (discussed later) which has been estimated to occur between about 108 000 and 114 000 yr BP (Smith and Foster 1969). The two other clear reversals in inclination have been called the Biwa I event, between about 176 000 and 186 000 yr BP, and the Biwa II event,

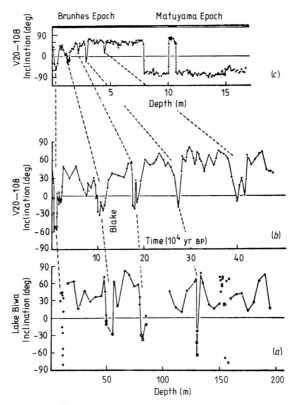

Figure 4.5   (*a*) Inclination against depth in core from Lake Biwa, Japan. (*b*) Inclination against time in deep-sea core V20-108 from the North Pacific (after Wollin *et al* 1971). (*c*) Inclination against depth in core V20-108 (after Ninkovich *et al* 1966, Nakajima *et al* 1973).

between 292 000 and 298 000 yr BP (Kawai *et al* 1972). Hayashida (1980) has confirmed the existence of the Biwa I excursion which he found in the Takashina Formation of the Kobiwako Group. The Kobiwako Group is considered to consist of the same sediments as those of the ancient Lake Biwa. Two volcanic ash layers in the Takashina Formation showed reversed polarity in an otherwise normal polarity section and were correlated in age with the Biwa I excursion seen in the lake core. A short reversal about the time of the Biwa II excursion has been observed in deep-sea cores in the north-western Pacific (Harrison 1974).

No clear reversals were seen in the upper part of the core although two very short excursions were seen at depths of 13 m and 26 m corresponding to times of approximately 18 000 and 49 000 yr BP. From the similarity of the pole paths Nakajima *et al* (1973) suggested that the earliest excursion may be the same as the Lake Mungo excursion. Nakajima *et al* placed the second excursion at around 38 000 yr BP and speculated that it may be the same as the Laschamp excursion. Hirooka (1976) found an excursion near Ina City, about 200 km from Lake Biwa, in sediments composed of layers of pumice, scoria, volcanic ash and fine grained sediments dated between about 60 000–35 000 yr BP. The VGP path is very similar to that of the second excursion seen at Lake Biwa around 49 000 yr BP.

A regional excursion (the Erieau excursion) has been proposed by Creer *et al* (1976b) on the basis of their study of two cores, nineteen kilometres apart, of late Wisconsinan sediments in Lake Erie. This excursion was dated indirectly as starting at about 14 000 yr BP and ending at a horizon whose age is somewhere between 10 500 and 7600 yr BP. The excursion manifests itself as a change in inclination from about $-90°$ to $+90°$. They suggested that the event could be the same as that reported by Yaskawa *et al* (1973) in the core from Lake Biwa (estimated age approximately 18 000 yr BP) and seen by Clark and Kennett (1973) in cores from the Gulf of Mexico, dated between 12 500 and 17 000 yr BP.

Banerjee *et al* (1979) obtained high resolution data of variations in the inclination recorded in the sediments of two post-glacial lakes in Minnesota. The first, from Lake St Croix, covered the period 0–9600 yr BP and the second, from Kylen Lake, from about 9000–16 000 yr BP. They found no evidence for any abnormal behaviour or excursions of the geomagnetic field over the past 16 000 yr in Minnesota, although several authors (e.g. Creer *et al* 1976b) have claimed that there have been sharp excursions in field direction during this time interval. Verosub and Banerjee (1977) have suggested that only a few of these proposed excursions represent real geomagnetic fluctuations; the rest, especially those recorded in sediments, probably being caused by disturbance of the sediment during its recovery or by imperfect recording of the geomagnetic field during deposition.

Vilks *et al* (1977) found zones of stable shallow geomagnetic inclination close to $0°$ (in contrast with the dipole value of $80°$) in two cores, 60 km apart, in the continental slope of the south-eastern Beaufort Sea. The magnetic 'event' took place between 10 000 and 40 000 yr BP—the authors prefer the latter date. Even though no reversals were seen, the authors believe that the anomalous palaeomagnetic inclination zones found in the cores represent real geomagnetic events.

Peirce and Clark (1978) found two reversely magnetised lava flows which were erupted subglacially near Maelifell in south-western Iceland, confirming an earlier report by Einarsson (1976). There is no firm age for the excursion—it is suggested that the eruptions occurred near the end of the Wisconsinan Ice Age, about 20 000 years ago. No radiometric dates are available to confirm this. They suggested that the excursion may be contemporaneous with the Laschamp and Lake Mungo events. They further pointed out that if the geomagnetic field reversed recently, one would expect to find evidence of a large increase in atmospheric ^{14}C due to increased production rates during the periods of low field strength associated with polarity transitions. If the transitions were 2000 years long, the increase would be about 25%. Longer transitions imply larger increases. However, a recent calibration of the radiocarbon timescale against $^{230}Th/^{234}U$ ages as old as 32 000 yr (Peng *et al* 1978, Stuiver 1978) indicates an increase of no more than 8%. If this calibration is correct, it implies either that the Maelifell event occurred at least 40 000 years ago, that the transitions were much more rapid than 2000 yr or that there was no major worldwide reduction in geomagnetic intensity associated with the event. Reversely magnetised basalts have also been found in the axial valley of the Mid-Atlantic Ridge (Ade-Hall *et al* 1973). The presence of the reversed outcrop at Maelifell, an area which is very similar tectonically to the Mid-Atlantic Ridge, suggests that similar reversed units there cooled during this reversed geomagnetic event. However, as this event has so far been discovered at only one locality in Iceland, its duration was probably quite short Peirce and Clark (1978) raise the very interesting question as to why the reversal has been observed so frequently on the Mid-Atlantic Ridge. They suggest that perhaps very young lavas (50 000 yr old) form a much greater part of the volcanism in the axial valley than earlier mapping has indicated.

Liddicoat *et al* (1979) carried out a palaeomagnetic investigation of marsh and near-shore lacustrine sediments deposited between about 25 000–5000 yr BP at Tlapacoya, Mexico. Normal polarity was found in all samples measured, although anomalous directions were found at one site in a mud unit dated about 14 500 yr BP. Subsequent work at six additional sites in the unit showed no abnormal directions. Liddicoat *et al* concluded that the anomalous directions are most likely not caused by an excursion of the magnetic field, although they could offer no convincing alternative explanation.

Abrahamsen and Knudsen (1979) carried out palaeomagnetic measurements on a single piston core in marine clay at Rubjerg, Denmark. They found an apparent excursion of the geomagnetic field in the undisturbed older marine deposits (the older Yolida Clay). They obtained seventeen specimens which exhibited the excursion in a section at least 1.2 m thick. The age is not well determined—foraminifera fauna and stratigraphic position indicate an age between 23 000 and 40 000 yr BP. They speculated that it may be correlated with the excursions reported at Laschamp, Mono Lake, Lake Mungo and in the Gulf of Mexico. Further work in Denmark has been carried out by Abrahamsen and Readman (1980) in clays at Nørre Lyngly. In the older Yolida Clay they confirmed the existence of the excursion seen in Rubjerg. In addition they found an excursion in declination in the Younger Yolida Clay, aged about 14 000 yr BP. This age was obtained from foraminifera fauna and by ^{14}C dating of *in situ* shells. They speculated that this excursion may be the same as that observed in the Gulf of Mexico (Clark and Kennett 1973) and in Lake Michigan (Dodson *et al* 1977).

Kristjansson and Gudmundsson (1980) found evidence for a geomagnetic reversal in lavas from three different hills in the active volcanic zone of the Reykjanes peninsula, south-west Iceland. The hills are thought to be contemporaneous and of late Wisconsinan age (12–13 000 yr BP). They considered it impossible that self-reversal has occurred in all these basalt units, as they cover a range of thicknesses, porosities and phenocryst contents, including in Skalamaelifell one highly oxidised (reddened) flow and three thin, subvertical dykes in tuff. Because of the short interval represented by the three lava sequences at Reykjanes, correlations with other geomagnetic excursions are bound to be very tentative. However, the mean magnetic field direction is similar to that occurring in the transitional lava outcrop near Maelifell reported by Peirce and Clark (1978). It is also consistent with the shallow geomagnetic inclinations found in Danish marine clays between 23 000 and 40 000 yr BP by Abrahamsen and Knudsen (1979).

A transition in the magnetic field which may be an excursion or a true reversal is the Blake event which was first discovered in seven cores from the Caribbean and Indian Ocean by Smith and Foster (1969) who dated it at between 105 000 and 114 000 yr BP and estimated its duration as 5000–7000 yr from their most reliable sedimentation rates, though the sedimentation rate in one core indicates a duration of 12 000 yr. The nature of the transition is primarily a change in sign in the inclination, and it is not clear that the excursion involved a full 180° rotation of the geomagnetic axis. Denham (1976) and Denham *et al* (1977) observed the Blake event in two large-diameter cores from the Greater Antilles outer ridge and reported that average sedimentation rates gave durations of 16 500 yr and 30 000 yr while another method led to an estimated duration of 50 000 yr. Denham did not consider such long durations to be realistic, since the Blake Event has been observed so rarely. He suggested that the VGP shift is consistent with a regional excursion caused by fluctuating non-dipole field intensities. Because of its apparent long duration the Blake Event may be distinct from other excursions. Sasajima *et al* (1980) found an excursion in pyroclastic flows at Keno and Kogashira, Japan and also in a rhyolitic tuff from Sigulagava, North Sumatra which they tentatively correlated with the Blake Event. If substantiated by further work this is the first time the Blake Event has been observed in igneous, continental rocks. Creer *et al* (1980) carried out palaeomagnetic and palaeontological studies on a long core from Gioia Tauro, Italy and claimed to have seen the Blake Event.

As recorded by the inclination, the Blake Event is split into two parts of estimated duration 16 000 yr and 23 000 yr separated by a short sequence of normal inclination of estimated duration 10 000 yr (see figure 4.6). This is discussed further in §7.2. Manabe (1977) claimed to have seen the Blake Event in a marine terrace in Japan, the dating of which is based primarily on palaeoclimatological evidence. Like the Italian record and two of the western North Atlantic records, the Blake Event in the Japanese record appears as two reversed intervals separated by a short normal interval.

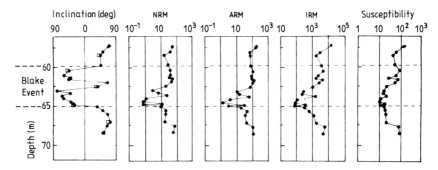

Figure 4.6   Logs of inclination and intensities of natural (NRM), anhysteretic (ARM) and isothermal (IRM) remanent magnetisations in μG, all after cleaning in 200 Oe AF and susceptibility in μG Oe^{-1} for samples recording the Blake Event. Open symbols on the inclination log show values before demagnetisation (after Creer *et al* 1980).

Another possible excursion of long duration is the Emperor event first reported by Ryan (1972) in cores from the Caribbean and Pacific with an age of between about 0.4–0.5 Myr. Later Wilson and Hey (1981) collected a series of marine magnetic profiles across the Galapagos spreading centre and found consistent short-wavelength anomalies which they interpreted as the Emperor event at 0.49 ± 0.05 Myr with a duration of about 10 000 yr. Champion *et al* (1981) also found evidence for a brief (5–10 000 yr) reversal of the field 0.46 ± 0.05 Myr ago, using cores through a sequence of basalt flows in the eastern Snake River Plain, Idaho. They suggested that it is the same as the Emperor event of Ryan and the Ureki event of Zubakov and Kochegura (1976). Wilson and Hey also suggested that this event was probably a global phenomenon—it is the youngest reversal documented by three independent techniques. Whether it should be classified as an excursion, however, is not clear.

Cox (1968) proposed that the frequency distribution of polarity epochs and events was continuous and that there were numerous undiscovered events with durations shorter than 50 000 yr (see §5.5). The ages of the principal polarity epoch boundaries for the last 5 Myr are now well determined, as are the times and durations of the major events (Mankinen and Dalrymple 1979). Discovery and confirmation of very short events, however, has been slow because they are of such short duration that evidence for their existence is not widely observed. Five short reversed events have been postulated in the Brunhes normal polarity epoch (see figure 4.7). All have been found, with varying frequency, in sediment cores, but until 1981 only the Laschamp event has been found in dated subaerial lava flows, and until the work of Wilson and Hey (1981) none had been reported in marine magnetic anomalies.

Two distinct normal polarity events have now been widely accepted in the Matuyama reversed polarity epoch. The more recent is the Jaramillo event; the other was originally called the Gilsa event, but is probably the same as that found at the Olduvai Gorge in Tanzania, Africa and named the Olduvai event (see §6.1 for a more detailed discussion). It has been suggested that short geomagnetic events or excursions exist in the Matuyama epoch in addition to these two well defined events. MacDougal and Chamalaun (1966) reported the possible occurrence of a short event around 2 Myr ago based upon palaeomagnetic and K–Ar age investigations of igneous rocks from Reunion Island. They also claimed that a very short reversed episode occurred at a late stage of the Olduvai event which would thus be split into two normal events. Other short polarity episodes have been found in deep-sea cores but most of them have not been considered seriously because of a lack of consistency between independent data.

Sueishi *et al* (1979) found two short normal polarity events in the Matuyama epoch in two deep-sea cores taken in the western equatorial Pacific. One is dated at about 1.07 Myr and the other at about 1.94 Myr. Both were accompanied by a pronounced drop in field intensity. The age of the later one is in agreement with that of a short reversal in declination seen by Ninkovich *et al* (1966) in a core from the North Pacific and by Watkins (1968) in four cores from the Southern Ocean. Sueishi *et al* believe that the observed 'reversal' in palaeomagnetic direction is not due to a polarity event but is more likely to be related to an excursion of the pole, because the observed declination fluctuates very rapidly within this core. They suggested that the most plausible explanation is that the moment of the Earth's dipole field decreased so much during this episode that palaeomagnetic directions, including repeated reversals, observed at various sites are mainly determined by the non-dipole field. The age of the earlier episode is in rough agreement with that of the Reunion event. The length of the transition is about 5000 yr, which would seem to classify it as an excursion. Heller and Liu (1982) examined a long core from a borehole near Lochuan, Shaanxi Province, China which penetrated a 136 m thick loess deposit. After thermal demagnetisation at 350 °C, several clearly

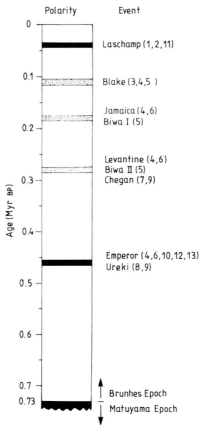

Figure 4.7 Postulated reversed polarity events within the Brunhes normal geomagnetic polarity epoch. Events recorded in dated lava flows are shown by solid shading; those found only in sediment cores by stippled patterns. Numbers refer to the following published reports: 1 Bonhommet and Babkine (1967), 2 Bonhommet and Zahringer (1969), 3 Smith and Foster (1969), 4 Wollin *et al* (1971), 5 Kawai *et al* (1972), 6 Ryan (1972), 7 Zubakov (1974), 8 Zubakov and Kochegura (1975), 9 Kochegura and Zubakov (1978), 10 Champion *et al* (1979), 11 Gillot *et al* (1979), 12 Champion *et al* (1981), 13 Wilson and Hey (1981) (after Champion *et al* 1981).

defined polarity zones were observed which they correlated with the Brunhes–Matuyama boundary and the Jaramillo and Olduvai events. No definite evidence of shorter term geomagnetic excursions was found (three samples out of a total of 231 do not fit the polarity sequence—two of them would fit the Reunion events). Intensity variations were especially pronounced in the upper part of the core (the last 1.2 Myr) and are interpreted to record rather drastic climatic changes coinciding with major episodes of loess deposition. Heller and Liu believe that the rock magnetic properties are largely controlled by chemical alteration during rock formation reflecting climatic fluctuations during late Pliocene to Pleistocene times.

Bingham and Stone (1976) examined 36 late Tertiary lava flows (3–4 Myr) from three separate sequences from the Wrangell Mountains of south-central Alaska. 27 out of the 36 flows gave mean directions more than 35° from an axial dipole or present field direction, suggesting that they were extruded during an excursion of the geomagnetic field. The data indicate that there were two different excursions around 3.4 Myr ago (since this age is near the Gilbert–Gauss boundary, the field behaviour may have occurred during polarity transitions). Since the probability of sampling three separate sections and finding that two of them contain different field excursions is rather low, Bingham and Stone suggested that such excursions may have occurred more frequently at the time of the extrusion of these lavas than appears to be the case in Quaternary times. This conclusion however rests on the very small number of sites compared. Later Hamilton and Evans (1983) examined 50 flows during the last 6.5 Myr at Level Mountain, a composite volcano in northern British Columbia. Four of the flows yielded a tight group of divergent directions which they interpreted as being erupted during a geomagnetic excursion or polarity transition near the base of the Gauss epoch. This lends some support to the suggestion of Bingham and Stone that the geomagnetic field was abnormally disturbed about 3.5 Myr ago—at least locally in north-western North America.

Steiner (1983) examined two deep-sea cores from the western Pacific containing the reversed polarity interval bounding the younger end of the Cretaceous long interval of normal polarity. Near the younger end of the reversal interval, an excursion of the inclination was found in both cores. No change in declination accompanied the change in inclination. Because of the stability of the samples to both AF and thermal demagnetisation and the fact that the change in sign of the inclination occurred at the same depth in sediments of the same age within two cores 500 m apart, Steiner believes that the cores contain a true record of the behaviour of the geomagnetic field. Approximate sedimentation rates indicate that the excursion had a duration of between 46 000 and 54 000 yr. Because of this rather long duration and its proximity (about 236 000–303 000 yr) to a field reversal, Steiner tentatively suggests that the excursion may have been an aborted attempt of the field to reverse.

Turner and Vaughan (1977) examined a core drilled off the north-east coast of England covering a complete sequence of the Marl Slate, and obtained evidence of rapid changes in the Permian magnetic field during the Zechstein marine transgression. During most of the Permian, the magnetic field was almost always reversed with very few cases of normal polarity. In one section of the core Turner and Vaughan found a transition from normal polarity to reversed polarity—the transition from positive to negative inclination was extremely short, although there was no sharp transition in declination. Both inclination and declination showed much less variation in the upper, negative part of the core. The authors believe that the magnetisation of the Marl Slate was acquired during deposition or early diagenesis, and that the lower positive zone is not a simple normal polarity zone, but rather evidence of rapid field changes similar to an excursion.

Bingham and Evans (1975) found evidence for a reversal and an earlier apparent excursion recorded in red siltstones of Precambrian age (approximately 1650–1800 Myr BP) in the Stark formation, Great Slave Lake, Canada. The excursion is seen in a five metre section approximately 150 m below the reversal, the palaeopole moving through about 80° from the stable direction above and below it. It is interesting that the path taken by the VGP during the excursion is close to that described by the later reversal, although there is a significant time difference between the two occurrences. This lends some support to the hypothesis that excursions represent aborted reversals. Hoffman (1981) has considered the question in some detail. In order to exclude any spurious or inaccurate data, he did not consider any excursion data obtained from sediments because of the controversies and uncertainties surrounding many sediment-recorded excursions (Verosub and Banerjee 1977, Verosub 1975, 1982). Similarly, excursion data obtained from sites at high latitude were not considered. Exclusion of Icelandic data is particularly justified, given the relatively close proximity of the recording site to the geographical pole. Hoffman found four excursion records which satisfied these stringent conditions, each within a sequence of basalt lavas ranging from mid-Tertiary to Pleistocene in age. Two of the four records occurred during the normal polarity of the field, and three of the recordings were obtained from sites in the southern hemisphere. In every case the VGP paths were either near-sided or far-sided consistent with transitional field geometries in which non-dipole axisymmetric components predominate (see §3.4). Independent of the hemisphere in which the records were obtained, reverse polarity excursions and R→N transitions were associated with near-sided VGP behaviour whereas normal polarity excursions and N→R transitions were associated with far-sided VGP behaviour. These findings, listed by Hoffman and Fuller (1978) as one possible regime, correspond to a predominantly quadrupolar (g_2^0) intermediate field geometry. Hoffman (1981) concluded that at least some excursions are aborted reversals.

4.7 Further comments

Although records have been obtained from a wide variety of environments and places, in no case has it yet been shown that a proposed excursion has been recorded with the same palaeomagnetic signature in two nearby but distinct environments. Again in no case has it been shown that two independently dated proposed excursions at widely separated sites are of precisely the same age. Demonstration of such a temporal consistency is critical in establishing a global as opposed to a regional nature of an excursion. Most geomagnetic excursions have been recorded in lake or deep-sea sediments and it is extremely difficult to assess the reliability of ages assigned to them. The horizons are usually dated by assuming uniform sedimentation rates between or beyond ^{14}C dated horizons. Varve counting techniques as used in dating the excursions recorded in Sweden (Noel and Tarling 1975) are generally regarded as being more reliable than ^{14}C dating. The problem of interpreting the data depends largely upon the extent to which different ages may be taken as being significantly different from one another.

Barbetti and Flude (1979b) have reviewed all available data on the strength of the Earth's magnetic field during the last 50 000 yr—the field appears to have been weaker than it is today for much of the period, 50 000–10 000 yr ago. They also discussed what effect this lowered field intensity would have had on the production of ^{14}C in the atmosphere. Some of the charged particle flux from interstellar space and the Sun is deflected by the Earth's magnetic field and thereby prevented from interacting with the atmosphere. The observed variation in geomagnetic field strength will thus have modulated that flux during the late Pleistocene, and Barbetti and Flude have estimated the effect on the ^{14}C timescale. Their conclusions differ slightly from those of Stuiver (1978), mainly in that they admit the possibility of errors greater than 2000 yr—errors up to about 5000 yr being not inconsistent with the data. Radiocarbon dates are also often affected by the presence of recycled fossil carbon which may have been deposited into the sediment as fragments of fossil shell or plant remains ('old carbon' effect) or incorporated directly into organisms living at the time of deposition from calcium carbonate originating from pre-existing limestones dissolved in the lake water ('hard water' effect). Both these effects can produce radiocarbon ages which are too old by up to several thousand years.

True reversals caused by a reversal of the dipole field should be capable of being seen worldwide, provided measurements in the correct time interval can be made. Excursions of the field could be caused by the main dipole field tilting at a large angle to the rotation axis and then returning to the same orientation as before. Such excursions should also be capable of being seen worldwide and the VGP paths from different locations should agree. The presence of the non-dipole field could allow excursions of the field to be seen at one location which are not observed at other locations far away, provided that the sources of the non-dipole field are sufficiently large, and provided that they are not located close to the centre of the Earth.

Marino and Ellwood (1978) have provided additional evidence that some apparent geomagnetic field excursions recorded in sediments may not represent short-term changes in the Earth's magnetic field. An excursion had been reported from Imuruk Lake, Alaska (Noltimier and Colinvaux 1976) dated at some 18 000 yr BP (^{14}C dates span 10 000–25 000 yr BP). Marino and Ellwood (1978) measured the magnetic fabric of both the excursion and normal sediments in the core using the magnetic susceptibility anisotropy method. They found that normal sediments exhibited a characteristic primary magnetic fabric, whereas excursion sediments exhibited an anomalous, distorted fabric. They suggested several causes for the distorted fabric such as slumping, deposition by turbidity currents, secondary mineralisation and increase in biogenic activity. Since the magnetic fabric of the excursion is anomalous, the magnetic directions recorded may be suspect and they suggested that without independent corroborative evidence, such excursions should be regarded with suspicion.

Løvlie and Holtedahl (1980) have given a further example of an 'apparent' excursion—an almost exact 180° change in declination being observed in a piston core of 90 cm core depth from the continental margin off the east coast of Norway. This inversion in declination is associated with changes in the inclination from a tightly grouped low dipping distribution above this depth to a more scattered distribution, broadly coinciding with the expected inclination of the present geomagnetic field, below 90 cm. The over-consolidated top section, defined by anomalously low inclination directions, shows a magnetic fabric typical of slurries or deposits affected by water currents, as opposed to an anomalous fabric in the lower section associated with magnetic directions coinciding with the present geomagnetic field. The sampling area shows seismic features indicative of slumping, erosion and redeposition, which are also reflected by micropalaeontological evidence. They concluded that the pre-consolidated top section has been transported by slumping, the anomalously low dipping directions resulting from processes acting during the initial consolidation. The effect of post-depositional compaction on the acquisition of a remanent magnetisation in fine-grained sediments has been experimentally shown to induce a reduction of the inclination (Blow and Hamilton 1978), which is comparable in magnitude to the decrease observed in the pre-consolidated section, assuming a Quaternary age for the latter. Løvlie and Holtedahl tentatively concluded that the anomalous direction of magnetisation reflects processes acting during consolidation. As compactive DRM is likely to play an important role in sediments overlain by ice during glacial periods (Blow and Hamilton 1975), the reported indication of a low inclination excursion at Rubjerg (Abrahamsen and Knudsen 1979) may possibly also be related to this effect.

Laboratory experiments have shown that misalignments affecting inclination and declination can be caused by the deposition process itself, by water currents and by the slope of the bedding plane. These errors may be present to varying degrees in different sedimentary environments (for a review, see Verosub 1977b). However, the maximum deviation which can arise is generally less than 30°, and cannot account for the large-scale anomalous directions that are sometimes observed.

As discussed in §2.1, studies of deep-sea sediments have shown that, from a palaeomagnetic point of view, there are 'stable' and 'unstable' cores. In an unstable deep-sea core, as much as 30–70% of the magnetic directions are anomalous so that the core may really consist of thin layers of stable and unstable materials. The boundaries between these layers may not show up as distinct lithological or chemical changes; rather, they may represent changes in mineralogy in a self-reversal region or a viscous magnetisation region. If a core contains only one layer or a few unstable layers the anomalous directions in these layers will have the characteristics of palaeomagnetic excursions. No palaeomagnetic excursion can be based on the evidence from a single core or site. At the very least there must be 'internal consistency' whereby the proposed palaeomagnetic excursion is found repeatedly in sediments of the same age in a given sedimentary environment. Bioturbation of the sea floor can also degrade and even erase portions of the palaeomagnetic record. Thus a regional geomagnetic fluctuation may not be recorded with the same signature at every site and may even be erased at some. Failure to detect a palaeomagnetic excursion in an adjacent lake or ocean basin is in itself an important result.

Finally, proposed palaeomagnetic excursions must show 'temporal consistency'. An event which only lasted a few hundred years cannot be recorded in two well dated cores at horizons which are several thousand years apart. In this regard, one must beware of the reinforcement syndrome (Watkins 1972): 'The initial report of a palaeomagnetic excursion will encourage other workers to re-examine previously unexplained or disregarded "curious" results and to reinterpret sedimentation rates so that the anomalous behaviour seen by them is contemporaneous with the palaeomagnetic excursions. Subsequent work will also focus on sediments of the same age. Reported excursions will then tend to cluster around a single date, whereas negative results showing no anomalous behaviour will tend to remain unpublished because they are not interesting.'

References

Abrahamsen N 1982 Magnetostratigraphy *The Pleistocene/Holocene boundary in south-western Sweden* ed O Olausson (Uppsala: Sver. Geol. Undersok)

Abrahamsen N and Knudsen K L 1979 Indication of a geomagnetic excursion in supposed Middle Weichselian interstadial marine clay at Rubjerg, Denmark *Phys. Earth Planet. Int.* **18** 238

Abrahamsen N and Readman P W 1980 Geomagnetic variations recorded in older ($\geqslant 23\,000$ BP) and younger Yolida Clay ($\simeq 14\,000$ BP) at Nørre Lyngly, Denmark *Geophys. J.* **62** 329

Ade-Hall J, Aumento F, Ryall P J C and Gerstein R E 1973 The mid Atlantic ridge near 45°N, XXI. Magnetic results from basalt drill cores from the Median Valley *Can. J. Earth Sci.* **10** 679

Banerjee S K, Lund S P and Levi S 1979 Geomagnetic record in Minnesota Lake sediments—absence of the Gothenburg and Eriau excursions *Geology* **7** 588

Barbetti M and Flude K 1979a Palaeomagnetic field strengths from sediments baked by lava flows of the Chaîne des Puys, France *Nature* **278** 153

——1979b Geomagnetic variation during the late Pleistocene period and changes in the radiocarbon timescale *Nature* **279** 202

Barbetti M F and McElhinny M W 1972 Evidence of a geomagnetic excursion 30 000 yr BP *Nature* **239** 327

——1976 The Lake Mungo geomagnetic excursion *Phil. Trans. R. Soc.* A **281** 515

Barbetti M, Taborin Y, Schmider B and Flude K 1980 Archaeomagnetic results from late Pleistocene hearths at Étiolles and Marsangy, France *Archaeometry* **22** 25

Barton C E and Polach H A 1971 ^{14}C ages and magnetic stratigraphy in three Australian maars *Radiocarbon* **22** 728

Bingham D K and Evans M E 1975 Precambrian geomagnetic field reversal *Nature* **253** 332

Bingham D K and Stone D B 1976 Evidence for geomagnetic field excursions and secular variation from the Wrangell Volcanoes of Alaska *Can. J. Earth Sci.* **13** 547

Blow R A and Hamilton N 1975 Palaeomagnetic evidence from DSDP cores of northward drift of India *Nature* **257** 570

——1978 Effect of compaction on the acquisition of a detrital remanent magnetisation in fine-grained sediments *Geophys. J.* **52** 13

Bonhommet N and Babkine J 1967 Sur la présence d'aimantations inversées dans la Chaîne des Puys *C.R. Acad. Sci. Paris* **264** 92

Bonhommet N and Zahringer J 1969 Palaeomagnetism and potassium–argon age determination of the Laschamp geomagnetic polarity event *Earth Planet. Sci. Lett.* **6** 43

Champion D E, Dalrymple G B, Kuntz M and Doherty D 1979 Reversed polarity lava flows within a Late Pleistocene volcanic sequence from Snake River Plain, Idaho: a possible reversed event within the Brunhes normal polarity epoch *Trans. Am. Geophys. Union* **60** 814

Champion D E, Dalrymple, G B and Kuntz M A 1981 Radiometric and palaeomagnetic evidence for the Emperor reversed polarity event at 0.46 ± 0.005 Myr in basalt lava flows from the Eastern Snake River plain, Idaho *Geophys. Res. Lett.* **8** 1055

Clark H C and Kennett J P 1973 Palaeomagnetic excursion recorded in latest Pleistocene deep-sea sediments, Gulf of Mexico *Earth Planet. Sci. Lett.* **19** 267

Coe R S 1977 Source models to account for Lake Mungo palaeomagnetic excursion and their implications *Nature* **269** 49

Condomines M 1978 Age of Olby–Laschamp geomagnetic polarity event *Nature* **276** 257

Cox A 1968 Length of geomagnetic polarity intervals *J. Geophys. Res.* **73** 3247

Creer K M, Anderson T W and Lewis C F M 1976a Late Quaternary geomagnetic stratigraphy recorded in Lake Erie sediments *Earth Planet. Sci. Lett.* **31** 37

Creer K M, Gross D L and Lineback J A 1976b Origin of regional geomagnetic variations recorded by Wisconsinan and Holocene sediments from Lake Michigan, USA and Lake Windermere, England *Geol. Soc. Am. Bull.* **87** 531

Creer K M, Readman P W and Jacobs A M 1980 Palaeomagnetic and palaeontological dating of a section at Gioia Tauro, Italy: identification of the Blake event *Earth Planet. Sci. Lett.* **50** 289

Denham C R 1974 Counter-clockwise motion of palaeomagnetic directions 24 000 years ago at Mono Lake, California *J. Geomag. Geoelec.* **26** 487

——1976 Blake polarity episode in two cores from the greater Antilles outer ridge *Earth Planet. Sci. Lett.* **29** 422

Denham C, Anderson R F and Bacon M P 1977 Palaeomagnetism and radiochemical age estimates for the Brunhes polarity episodes *Earth Planet. Sci. Lett.* **35** 384

Denham C R and Cox A 1971 Evidence that the Laschamp polarity event did not occur 13 300–30 400 years ago *Earth Planet. Sci. Lett.* **13** 181

Dodson R E, Fuller M D and Kean W F 1977 Palaeomagnetic records of secular variation from Lake Michigan sediment cores *Earth Planet. Sci. Lett.* **34** 387

Doh S J and Steele W K 1981 The late Pleistocene geomagnetic field as recorded by sediments from Fargher Lake, Washington *Trans. Am. Geophys. Union* **62** 851

——1983 The late Pleistocene geomagnetic field as recorded by sediments from Fargher Lake, Washington, USA *Earth Planet. Sci. Lett.* **63** 385

Einarsson T 1976 *Soc. Sci. Isl. Greinar* **5** 119

Freed W K and Healy N 1974 Excursions of the Pleistocene geomagnetic field recorded in Gulf of Mexico sediments *Earth Planet. Sci. Lett.* **24** 99

Gillot P Y, Labeyrie J, Laj C, Valladas G, Guérin G, Poupeau G and Delibrias G 1979 Age of the Laschamp polarity palaeomagnetic excursion revisited *Earth Planet. Sci. Lett.* **42** 444

Guérin G, Gillot P-Y, Reyss J-L and Valladas G 1984 Datation par thermoluminescence et Potassium–Argon de coulées volcaniques récentes. Application à la Chaîne des Puys *Earth Planet. Sci. Lett.* submitted

Guérin G and Valladas G 1980 Thermoluminescence dating of volcanic plagioclases *Nature* **286** 697

Hall C M and York D 1978 K–Ar and ^{40}Ar/^{39}Ar age of the Laschamp geomagnetic polarity reversal *Nature* **274** 462

Hall C M, York D and Bonhommet M 1979 ^{40}Ar/^{39}Ar dating of the Laschamp event and associated volcanism in the Chaîne des Puys *Trans. Am. Geophys. Union* **60** 244

Hamilton T S and Evans M E 1983 A magnetostratigraphic and secular variation study of Level Mountain, northern British Columbia *Geophys. J.* **73** 39

Harrison C G A 1974 The palaeomagnetic record from deep-sea sediment cores *Earth Sci. Rev.* **10** 1

Harrison C G A and Ramirez E 1975 Areal coverage of spurious reversals of the Earth's magnetic field *J. Geomag. Geoelec.* **27** 139

Hayashida A 1980 Confirmation of a magnetic polarity episode in the Brunhes normal epoch; a preliminary report *Rock Magnetism and Palaeogeophys.* **7** 85

Heller F 1980 Self-reversal of natural remanent magnetisation in the Olby–Laschamp lavas *Nature* **284** 334

Heller F and Liu T-S 1982 Magnetostratigraphical dating of loess deposits in China *Nature* **300** 431

Heller F and Petersen N 1982 The Laschamp excursion *Phil. Trans. R. Soc.* A **306** 169

Heusser C J and Heusser L E 1980 Sequence of pumiceous tephra layers and the late Quaternary environmental record near Mount St Helens *Science* **210** 1007

Hirooka K 1976 Some notes on the characteristics of geomagnetic excursion in late

Pleistocene, in Japan *Palaeolimnology of Lake Biwa and the Japanese Pleistocene* vol 4 ed S Horie (Otsu: Kyoto University) p 153

Hoffman K A 1981 Palaeomagnetic excursions, aborted reversals and transitional fields *Nature* **294** 67

Hoffman K A and Fuller M 1978 Transitional field configuration and geomagnetic reversal *Nature* **273** 715

Huxtable J and Aitken M J 1977 Thermoluminescent dating of Lake Mungo geomagnetic polarity excursion *Nature* **265** 40

Huxtable J, Aitken M J and Bonhommet N 1978 Thermoluminescence dating of sediment baked by lava flows of the Chaîne des Puys *Nature* **275** 207

Johnson H G, Kinoshita H and Merrill R T 1975 Rock magnetism and palaeomagnetism of some north Pacific deep-sea sediment cores *Geol. Soc. Am. Bull.* **86** 412

Kawai N, Yaskawa K, Nakajima T, Torii M and Horie S 1972 Oscillating geomagnetic field with a recurring reversal discovered from Lake Biwa *Proc. Jap. Acad.* **48** 186

Kochegura V V and Zubakov V A 1978 Paleomagnetic timescale of the Ponto–Caspian Plio–Pleistocene deposits *Palaeogeog., Palaeoclimatol., Palaeoecol.* **23** 151

Kristjansson L and Gudmundsson A 1980 Geomagnetic excursion in late-glacial basalt outcrops in south-western Iceland *Geophys. Res. Lett.* **7** 337

Lajoie K R and Liddicoat J C 1980 Refinement of the chronology and palaeomagnetic record at Mono Lake, California *Trans. Am. Geophys. Union* **61** 215

Larson E E, Watson D E and Jennings W 1971 Regional comparison of a Miocene geomagnetic transition in Oregon and Nevada *Earth Planet. Sci. Lett.* **11** 391

Liddicoat J C and Coe R S 1979 Mono Lake geomagnetic excursion *J. Geophys. Res.* **84** 261

Liddicoat J C, Coe R S, Lambert P W and Valastro S Jr 1979 Palaeomagnetic record in Late Pleistocene and Holocene dry lake deposits at Tlapacoya, Mexico *Geophys. J.* **59** 367

Liddicoat J C, Lajoie K R and Sarna-Wojcicki A M 1982 Detection and dating of the Mono Lake excursion in the Lake Lahontan Sehoo Formation, Carson Sink, Nevada *Trans. Am. Geophys. Union* **63** 920

Løvlie R and Holtedahl H 1980 Apparent palaeomagnetic low-inclination excursion on a preconsolidated continental shelf sediment *Phys. Earth Planet. Int.* **22** 137

MacDougal I and Chamalaun F H 1966 Geomagnetic polarity scale of time *Nature* **212** 1415

Manabe K I 1977 Reversed magneto-zone in the Late Pleistocene sediments from the Pacific coast of Odaka, northeast Japan *Quat. Res.* **7** 372

Mankinen E A and Dalrymple G B 1979 Revised geomagnetic polarity time scale for the interval 0–5 Myr BP *J. Geophys. Res.* **84** 615

Marino R J and Ellwood B B 1978 Anomalous magnetic fabric in sediments which record an apparent geomagnetic field excursion *Nature* **274** 581

Morner N A 1976 Palaeomagnetism in deep-sea core A179-15: a reply *Earth Planet. Sci. Lett.* **29** 240

Morner N A 1977 The Gothenburg magnetic excursion *Quat. Res.* **7** 413

Morner N A and Lanser J P 1974 Gothenburg magnetic 'flip' *Nature* **251** 408

——1975 Palaeomagnetism in deep sea core A179-15 *Earth Planet. Sci. Lett.* **26** 121

Morner N A, Lanser J P and Hospers J 1971 Late Weichselian palaeomagnetic reversal *Nature Phys. Sci.* **234** 173

Nakajima T, Yaskawa K, Natsuhara N, Kawai N and Horie S 1973 Very short period geomagnetic excursion 18 000 yr BP *Nature Phys. Sci.* **244** 8

Ninkovich D, Opdyke N D, Heezen B C and Foster J H 1966 Palaeomagnetic stratigraphy, rates of deposition and tephrachronology in North Pacific deep-sea sediments *Earth Planet. Sci. Lett.* **1** 476

Noel M and Tarling D H 1975 The Laschamp geomagnetic 'event' *Nature* **253** 705

Noel M, Thompson R and Berglund B 1977 The late Weichselian geomagnetic event *Nature* **267** 181

Noltimier H C and Colinvaux P A 1976 Geomagnetic excursion from Imuruk Lake, Alaska *Nature* **259** 197

Oberg C J and Evans M E 1977 Spectral analysis of Quaternary palaeomagnetic data from British Columbia and its bearing on geomagnetic secular variation *Geophys. J.* **51** 691

Opdyke N D 1976 Discussion of paper by Morner and Lanser concerning the palaeomagnetism of deep sea core A179-15 *Earth Planet. Sci. Lett.* **29** 238

Opdyke N D, Shackleton N J and Hays J D 1974 The details of a magnetic excursion as seen in a piston core from the Southern Indian Ocean *Trans. Am. Geophys. Union* **55** 237

Palmer D F, Henyey T L and Dodson R E 1979 Palaeomagnetic and sedimentological studies at Lake Tahoe, California, Nevada *Earth Planet. Sci. Lett.* **46** 125

Peirce J W and Clark M J 1978 Evidence from Iceland on geomagnetic reversal during the Wisconsin Ice Age *Nature* **273** 456

Peng T H, Goddard J G and Broecker W C 1978 A direct comparison of ^{14}C and ^{230}Th ages at Searles Lake, California *Quat. Res.* **9** 319

Ransom C J 1973 Magnetism and Archaeology *Nature* **242** 518

Ryan W B F 1972 Stratigraphy of late Quaternary sediments in the eastern Mediterranean *The Mediterranean Sea: A Natural Sedimentation Laboratory* ed D J Stanley (Stroudsburg: Dowden, Hutchison & Ross)

Sasajima S, Nishimura S and Hirooka K 1980 Studies on the Blake episode with special emphasis to East Asian results obtained *Rock Magnetism and Palaeomagnetism* **7** 90

Smith D J and Foster J H 1969 Geomagnetic reversal in Brunhes normal polarity epoch *Science* **163** 565

Steiner M B 1983 Geomagnetic excursion in the late Cretaceous *Geophys. J.* **73** 17

Stuiver M 1978 Radiocarbon timescale tested against magnetic and other dating methods *Nature* **273** 271

Sueishi T, Sato T, Kawai N and Kobayashi K 1979 Short geomagnetic episodes in the Matuyama epoch *Phys. Earth Planet. Int.* **19** 1

Sukroo J C, Christofel D A, Vella P and Topping W W 1978 Rejecting evidence of Gothenburg geomagnetic reversal in New Zealand *Nature* **271** 650

Tanaka H and Tachibana K 1979 A geomagnetic reversal in the latest Brunhes epoch discovered at Shibutami, Japan *Rock magnetism and Palaeogeophys.* **6** 38

——1981 A geomagnetic reversal in the latest Brunhes epoch discovered at Shibutami, Japan *J. Geomag. Geoelec.* **33** 287

Thompson R and Berglund B 1976 Late Weichselian geomagnetic 'reversal' as a possible example of the reinforcement syndrome *Nature* **263** 490

Turner G M, Evans M E and Hussin I B 1982 A geomagnetic secular variation study (31 000–19 000 BP) in Western Canada *Geophys. J.* **71** 159

Turner P and Vaughan D J 1977 Evidence of rapid changes in the Permian geomagnetic field during the Zechstein marine transgression *Nature* **270** 593

Valladas G, Gillot P Y, Poupeau G and Reyss J L 1977 *5th Eur. Conf. Geochronology, Pisa*

Verosub K L 1975 Palaeomagnetic excursions as magnetostratigraphic horizons: a cautionary note *Science* **190** 48

——1977a The absence of the Mono Lake geomagnetic excursion from the palaeomagnetic record of Clear Lake, California *Earth Planet. Sci. Lett.* **36** 219

——1977b Depositional and post-depositional processes in the magnetisation of sediments *Rev. Geophys. Space Phys.* **15** 129

——1982 Geomagnetic excursions: a critical assessment of the evidence as recorded in sediments of the Brunhes epoch *Phil. Trans. R. Soc.* A **306** 161

Verosub K L and Banerjee S K 1977 Geomagnetic excursions and their palaeomagnetic record *Rev. Geophys. Space Phys.* **15** 145

Verosub K L, Davis J O and Valastro S Jr 1980 A palaeomagnetic record from Pyramid Lake, Nevada, and its implication for proposed geomagnetic excursions *Earth Planet. Sci. Lett.* **49** 141

Vilks G, Hall J M and Piper D J W 1977 The natural magnetisation of sediment cores from the Beaufort Sea *Can. J. Earth Sci.* **14** 2007

Watkins N D 1968 Short period geomagnetic polarity events in deep-sea sedimentary cores *Earth Planet. Sci. Lett.* **4** 341

Watkins N D 1972 Review of the development of the geomagnetic polarity timescale and discussion of prospects for its finer definition *Geol. Soc. Am. Bull.* **83** 551

Wilson R L, Dagley P and McCormack A G 1972 Palaeomagnetic evidence about the source of the geomagnetic field *Geophys. J.* **28** 213

Wilson D S and Hey R N 1981 The Galapagos axial magnetic anomaly: evidence for the Emperor reversal within the Brunhes and for a two-layer magnetic source *Geophys. Res. Lett.* **8** 1051

Wintle A G 1973 Anomalous fading of thermoluminescence in mineral samples *Nature* **245** 143

——1977 Thermoluminescence dating of minerals—traps for the unwary *J. Electrostatics* **3** 281

Wollin G, Ericson D B, Ryan W B F and Foster J M 1971 Magnetism of the Earth and climate changes *Earth Planet. Sci. Lett.* **12** 175

Yaskawa K 1974 Reversals, excursions and secular variations of the geomagnetic field in the Brunhes normal polarity epoch *Palaeolimnology of Lake Biwa and the Japanese Pleistocene* ed S Horie (Otsu: Kyoto University) **2** 77

Yaskawa K, Nakajima T, Kawai N, Torii M, Notsuhara N and Horie S 1973 Palaeomagnetism of a core from Lake Biwa (I) *J. Geomag. Geoelec.* **25** 447

Zubakov V A (ed) 1974 *Geokhronologia SSR* tom 3 (Noveishy etap) (Leningrad: Nedra)

Zubakov V A and Kochegura V V 1976 *Ob obratnoi namagnichennosti drevneeuxsinskikh sloev Urekskogo vazreza (Prichernomorje)* XXVII Gerzenovskie chtenya, Leningrad 31

5 Models for reversals

5.1 The disc dynamo

Because of the complexity of the non-linear equations that describe the geodynamo, various models have been proposed. One of the simplest analogies is the homopolar disc dynamo (see figure 5.1) first suggested and studied by Bullard (1955). It consists of an electrically conducting disc which can be made to rotate on an axle by an applied couple. If it rotates in an axial magnetic field, a radial EMF will be produced between the axle and the edge of the disc. If this were all, the EMF would be balanced by an electric charge on the edge of the disc and no current would flow. If one end of a stationary coil coaxial with the disc and axle is joined to the edge of the disc by a sliding contact (a brush) and the other end is joined to another sliding contact on the axle a current will flow through the coil and an axial magnetic field will be produced. No external source of field or current is required and no part of the machine is ferromagnetic. This system becomes a dynamo when this induced field becomes equal to the field required to produce it. It must not be forgotten, however, that the homopolar dynamo is not simply connected. It requires complicated conduction paths and slipping brushes in order to operate; and it is not at all obvious that a homogenous conducting fluid can be put into motion in such a way as to initiate dynamo action.

It is easier to formulate the equations for the system if the current in the disc is axially symmetric; this can be achieved by substituting a highly conducting ring for the brush. The current in the disc can then be represented by a single variable and is not a function of azimuth.

If the disc is driven by a constant couple G, the equation of motion is

$$C\dot{\omega} = G - MI^2 \tag{5.1}$$

where C is the moment of inertia of the disc, ω its angular velocity, I the current and $2\pi M$ the mutual inductance of the coil and disc. The equation governing the current is

$$L\dot{I} + RI = M\omega I \tag{5.2}$$

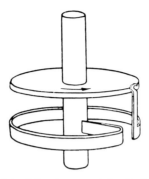

Figure 5.1   Disc dynamo (after Bullard 1955).

where L and R are the inductance and resistance of the coil. The solution of these equations has been discussed by Bullard (1955). There are two non-linear terms, I^2 in equation (5.1) and ωI in equation (5.2). Since equation (5.1) is quadratic in I and equation (5.2) is linear and homogeneous in I, the dynamo can produce a current (and therefore a magnetic field) in either direction. It cannot, however, switch from one direction to the other since, if I is zero, so is $\dot{I}$. In order to show reversals another term must be introduced into equation (5.2) which is not proportional to either I or $\dot{I}$. This can be achieved by using two coupled dynamos (Rikitake 1958), or by adding an impedance between the brush and the coil and a shunt connected across the coil of a single dynamo (Malkus 1972). This latter arrangement is shown in figure 5.2. If R_s and L_s and R_b and L_b are the resistance and inductance of the shunt and series impedance respectively and I_s the current carried by the shunt, the equations of the system are

$$C\dot{\omega} = G - MI(I + I_s) \tag{5.3}$$

$$L_s\dot{I}_s + R_sI_s = M\omega I \tag{5.4}$$

$$(L + L_b)\dot{I} + (R + R_b)I + L_b\dot{I}_s + R_bI_s = M\omega I. \tag{5.5}$$

The properties of these equations have been investigated by Robbins (1977) who showed that there are four possible regimes depending on the values of the different parameters. If $L/R > L_s/R_s$ and $L_b/R_b > L_s/R_s$ irregular reversing solutions occur. Robbins has studied this regime in some detail for the simplest case in which $L_s = 0$. She showed that there exist solutions whose path in phase space stays in the neighbourhood of one equilibrium point and then switches to the neighbourhood of the other without ever being captured by either. The currents in the coil and shunt show growing oscillations around one equilibrium state culminating in a reversal of current and a transition to oscillations about the other. An example of oscillations produced in this way is shown in figure 5.3. The number of oscillations between reversals is very variable—in the example shown it varied from 1 to 66. Other solutions exist

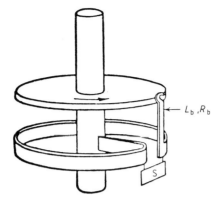

Figure 5.2   Disc dynamo with an impedance between the brush and the coil and a shunt connected across the coil (after Bullard 1978).

where the erratic behaviour does not persist and where, after 18 reversals, the solution is captured by one of the equilibrium points.

If we suppose that the fluid motion in the Earth's core consists of a number of eddies and that each eddy motion is represented by a disc, then we can imagine a series of disc dynamos each coupling with the next stage. The field produced by the last stage may be fed to the first stage. Rikitake (1958) discussed the case of two identical coupled disc dynamos in which the current from each disc energises the coil of the other (see figure 5.4). He found by numerical integration that it was possible for the currents to reverse sign. Later Allan (1958, 1962) showed that the reversals had an apparently random distribution in time. Further numerical and analytical investigations have been carried out by Mathews and Gardner (1963), Sommerville (1967) and Cook and Roberts (1970).

If the two dynamos are similar and the couples applied to them are equal, the equations are

$$L\dot{I}_1 + RI_1 = M\omega_1 I_2 \tag{5.6}$$

$$L\dot{I}_2 + RI_2 = M\omega_2 I_1 \tag{5.7}$$

$$C\dot{\omega}_1 = G - MI_1 I_2 \tag{5.8}$$

$$C\dot{\omega}_2 = G - MI_1 I_2 \tag{5.9}$$

where the suffixes 1 and 2 refer to the first and second dynamos. The equations can be put into non-dimensional form by measuring time t in units of $(CL/GM)^{1/2}$, currents in units of $(G/M)^{1/2}$ and angular velocities in units of $(GL/CM)^{1/2}$. Writing

$$I_i = (G/M)^{1/2} X_i \qquad \omega_i = (GL/CM)^{1/2} Y_i \qquad i = 1, 2 \tag{5.10}$$

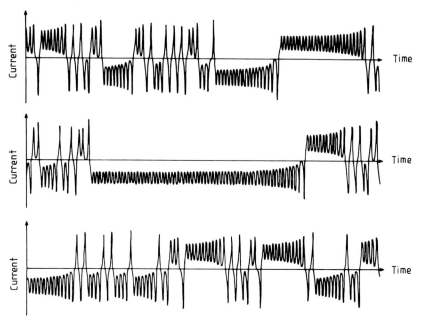

Figure 5.3   Oscillations of the current in the coil of the disc dynamo with shunt and series impedance (after Robbins 1977).

the governing equations (5.6)–(5.9) become

$$\dot{X}_1 + \mu X_1 = Y_1 X_2 \tag{5.11}$$

$$\dot{X}_2 + \mu X_2 = Y_2 X_1 \tag{5.12}$$

$$\dot{Y}_1 = \dot{Y}_2 = 1 - X_1 X_2 \tag{5.13}$$

where

$$\mu = (CR^2/GLM)^{1/2} \tag{5.14}$$

is the ratio of the mechanical acceleration time (in the absence of magnetic field) to the electromagnetic decay time (in the absence of motion). Equations (5.11)–(5.13) have, in general, no known analytic solutions.

It follows from equation (5.13) that $Y_1 - Y_2 = A$ (a constant), i.e. the difference in angular velocities is constant. Figure 5.5 illustrates the behaviour of the system for $\mu = 1, A = 3.75$ and clearly shows reversals taking place at time $t = 11, 21, 44, 51, 53, 71, 85$ etc.

Although it is clear that such two-disc dynamos are unstable and oscillate and reverse, it is very doubtful if any quantitative results have been obtained which are applicable to the Earth. The behaviour of such models depends very much on the value of μ. Both Allan (1962) and Lowes (unpublished) have

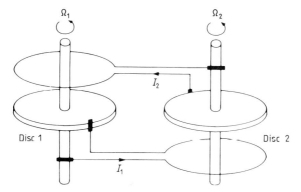

Figure 5.4 Coupled system of two disc dynamos (after Rikitake 1958).

estimated μ as lying between about 10^{-2} and 10^{-3} for the core of the Earth. All earlier published results were for values of μ of about unity. Because of computational difficulties no solutions have been obtained for values less than 10^{-2}, but all indications are that the solution would consist of periodic reversals of field with a period corresponding to only a few years in the case of the Earth.

Rikitake's treatment of two coupled disc dynamos has been generalised by Nozières (1978). He showed that the non-linear dynamical equations involve two very different timescales—a fast MHD scale and a slow inductive scale. For a certain range of the parameters involved the magnetic field suddenly reverses, the intervals between reversals being the slow timescale, and the duration of the reversal the fast timescale. Such a relaxation mechanism explains quantitatively the comparative rapidity of an actual reversal compared to the time interval between reversals. In his treatment Nozières allowed only one mechanical variable—the magnitude of the velocity field at any time, i.e. he assumed a constant geometry for the convection pattern. In Bullard's homopolar dynamo there is in addition to one mechanical variable (the angular velocity), one electrical variable (the current). Nozières broadened the scale of the problem by allowing two electromagnetic degrees of freedom. Using the ratio of the two timescales as an expansion parameter, he obtained an analytical solution of the non-linear differential equations which describe his model, and investigated the stability of the slow motion against fast MHD oscillations. Among the various possible regimes he found a 'ratchetlike' relaxation behaviour which may account for the very sudden reversals of the field. An overall cyclic slow motion is driven by the inductive forces. Periodically, the MHD fast oscillations become unstable, and the system jumps to a new equilibrium configuration, in which J and B are roughly reversed.

There exists a wide class of systems of ordinary differential equations which represent forced dissipative hydrodynamic systems such as geodynamos, which have non-periodic solutions. Such systems may oscillate randomly

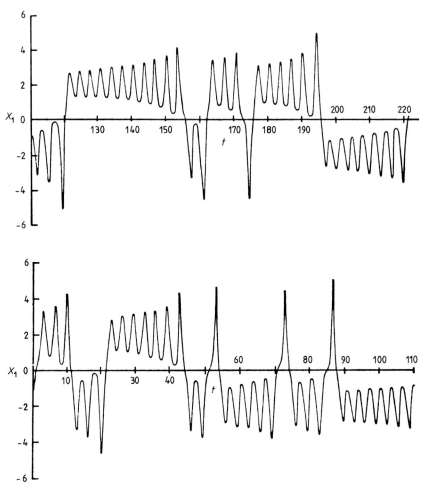

Figure 5.5   A typical evolution of the current X_1 as a function of time (in seconds) (after Cook and Roberts 1970).

between two states of fixed points; i.e. they may have two unstable polarity states. Rikitake's double-disc model belongs to such a class and such a model displays non-periodic reversals in the absence of any triggering mechanism. Lorenz (1963) first pointed out such behaviour in a discussion of the feasibility of very long range weather forecasting. The phenomenon (since named 'chaos') has been found in many other fields, e.g. models of population dynamics, physiological control systems and chemical reactions. The Rikitake system has two equilibrium points N and R $(\pm K, \pm K^{-1}, \mu K)$ in (X_1, X_2, Y_1) phase space where K is given by

$$A = \mu(K^2 - K^{-2}). \tag{5.15}$$

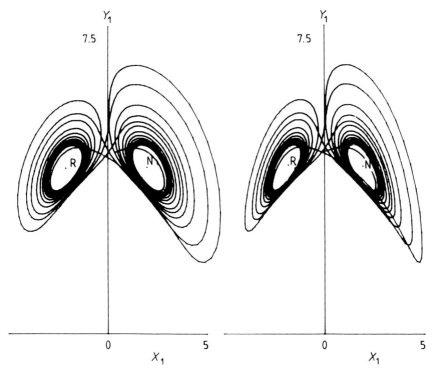

Figure 5.6   Stereo plot for an orbit for $\mu=1$ and $K=2$ in (X_1, X_2, Y_1) phase space. N and R are unstable fixed points. The orbit circles round N or R on an attracting plane which looks like two curved discs (after Ito 1980).

Both are unstable foci—around them there is an 'attracting' plane which traps all orbits starting from any point except those on the Y_1 axis. An orbit circles around N or R on this plane, irregularly travelling from an orbit round one point to one around the other (figure 5.6). This corresponds to a polarity reversal of the magnetic field. Ito (1980) has investigated the statistical properties of this system. Figure 5.7 is a phase diagram in the parameter space (μ, K) showing various regions of periodic regime $(P_1, P_2, \dots)$ and a chaotic regime. The transition from the periodic to the chaotic regime is characterised by a succession of period doubling bifurcations (see figure 5.8). Ito found in the centre of the chaotic regime a parameter region in which reversals seldom occur and the dynamics is less disordered. The Markov entropy of the Lorenz map for the system has a sharp minimum in this parameter region which Ito calls the minimum entropy regime. The smallness and non-uniformity of the frequency of reversals as shown by the palaeomagnetic data suggest that the present geodynamo is in such a state of minimum entropy.

Merrill *et al* had earlier suggested (1979) that hydromagnetic dynamos

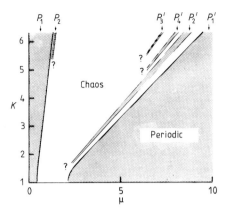

Figure 5.7   Phase diagram of the Rikitake system (μ, K) space. Solutions are periodic in stippled regions $(P_1, P_2, \ldots, P'_1, P'_2, \ldots)$, but are non-periodic in regions denoted as chaos. The sequence of $P_1, P_2, \ldots$ is supposed to be a mirror sequence of $P'_1, P'_2, \ldots$ (after Ito 1980).

might exist in which there are at least two different acceptable velocity fields and in which changes in polarity occur only when there is a change from one velocity field to the other. The existence of two velocity states suggests bifurcating solutions that are frequently encountered in fluid dynamical systems (e.g. Chandrasekhar 1961, Huppert and Moore 1976, Roberts 1978). A characteristic of many finite amplitude systems is that more than one state is stable or metastable for a given set of externally imposed conditions. It has been speculated (for example, by Roberts 1978) that reversals could be associated with 'catastrophic' transitions between primarily geostrophic and primarily magnetostrophic states. If this interpretation is correct, then it is perhaps surprising that the normal and reversed fields are apparently so similar.

In 1958 Herzenberg proposed a dynamo model of the Earth's magnetic field which consisted of two spheres in the Earth's core each of which rotates as a rigid body at a constant angular velocity about a fixed axis. The axially symmetric component of the magnetic field of one of the spheres is twisted by rotation resulting in a toroidal field which is strong enough to give rise to a magnetic field in the other sphere. The axial component of this field is twisted as well and fed to the first sphere. If the rotation of the spheres is sufficiently rapid, a steady state may be reached. Lowes and Wilkinson (1963) built a working model of what is effectively a homogeneous self-maintaining dynamo based on Herzenberg's theory. For mechanical convenience they used, instead of spheres, two cylinders placed side by side with their axes at right angles to one another so that the induced field of each is directed along the axis of the other. If the directions of rotation are chosen correctly, any applied field along one axis will lead, after two stages of induction, to a parallel induced field. If the

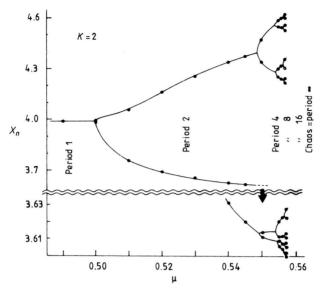

Figure 5.8 Bifurcation of stable fixed points in the Lorenz map as μ is increased in the transitional zone of the periodic regime P_1 at $K=2$. The ordinate X_n is the value of the nth maximum of $|X_1|$. The figure illustrates a cascade of period doubling bifurcations (after Ito 1980).

velocities are large enough the induced field will be larger than the applied field which is no longer needed, i.e. the system would be self-sustaining. Wilkinson has since carried out many more experiments using as many as four cylinders and has been able to produce reversing magnetic fields, the reversal characteristics depending on the experimental configuration used. In most cases the reversal waveform is periodic, unlike the reversal record of the Earth's magnetic field, and yields a simple spectrum when analysed into its Fourier components. In some arrangements a more complex reversal pattern develops and analyses of these records reveal complicated spectra (Kerridge and Wilkinson 1983).

5.2 Theoretical models

Two basic models have been suggested to describe the behaviour of the geomagnetic field during a polarity transition. In one the main dipole field remains axial but decreases in strength during the transition so that the non-dipole field becomes dominant in the intermediate stages (see e.g. Larson *et al* 1971). In the other the non-dipole field remains a small part of the total field but the dipole is considered to have two or three components, the strength, polarity and possible orientation of which change during the transition.

Bochev (1969) had earlier suggested such a model with three dipoles—for epoch 1960, he found that two dipoles are approximately parallel to the Earth's axis of rotation whilst the third is aligned obliquely to the axis. Creer and Ispir (1970) suggested that these three dipoles correspond to different dynamo processes and that each dynamo might drift or undergo periodic changes in position—each might also fluctuate in strength and even reverse its polarity independently of the others.

Verosub (1975a) made a similar suggestion to account for reversals. In his model the field arises from two separate sources, the observed dipole component of the Earth's field being the vector sum of the dipole components of each source, i.e.

$$M_0 = M_1 + M_2. \tag{5.16}$$

If the dipole components of each source are oppositely directed, one pointing essentially towards geographical north and the other south, then

$$M_0 \simeq M_1 - M_2. \tag{5.17}$$

The prevailing magnetic polarity reflects the polarity of the dominant source component: a magnetic reversal, which corresponds to a change in the sign of M_0, represents a shift in the relative sizes of M_1 and M_2. If both M_1 and M_2 are very large compared with M_0, then only a small fluctuation in M_1, M_2 or both will result in a reversal. If the Earth's field does arise from two sources, then differences could be expected between normal and reversed polarity states, indicating that a different source is dominant in each state. Some authors claim to have seen such effects. Wilson (1972) found significant differences in the time-averaged mean pole positions of normal and reversed populations while Dagley and Lawley (1974) found a clear predominance of reversed over normal transitions and concluded that the reversed state is less stable than the normal one. However as mentioned in §3.1 Kristjansson and McDougal (1982), as a result of more detailed studies, concluded that normal and reversely magnetised states of the geomagnetic field were equally probable during the last 14 Myr. They found complete equivalence between normal and reversely magnetised lavas (in Iceland) in all statistical properties of their intensity distribution.

Verosub (1975b) developed a new method of modelling the Earth's field by fitting a set of spherical harmonic coefficients with a geocentric axial dipole and an eccentric (but not necessarily axial) dipole. The free parameters are the relative strengths of the two dipoles and the coordinates of the eccentric dipole. When applied to the first eight coefficients of the International Geomagnetic Reference Field for 1965, the best fit consists, not of two more or less parallel components whose sum is the present field, but rather of two large, more or less antiparallel components whose difference is the Earth's field. This is precisely the configuration described by equation (5.17). Verosub suggested that the two sources are located in the inner and outer cores. The source field in the liquid outer core could be generated by MHD action, but it is much more difficult to postulate a source mechanism in the solid, inner core.

5.3 The models of Cox and Parker

Cox (1968) developed a probabilistic model in which he assumed that polarity changes occur as a result of an interaction between steady oscillations and random processes. The steady oscillator is the dipole component of the field and the random variations are the components of the non-dipole field. The random variations serve as a triggering mechanism that produces a reversal whenever the ratio of the non-dipole to dipole fields exceeds a critical value (see figure 5.9). Parker (1969) developed an alternative model for reversals. He showed that a fluctuation in the distribution of the cyclonic convective cells[†] in the core can produce an abrupt reversal of the geomagnetic field.

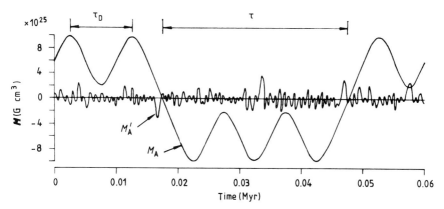

Figure 5.9   Probabilistic model for reversals. τ_D is the period of the dipole field and τ the length of a polarity interval. A polarity change occurs whenever the quantity $(M_A + M'_A)$ changes sign, where M_A is the axial moment of the dipole field and M'_A is a measure of the non-dipole field (after Cox 1968).

The simplest fluctuation leading to a reversal is a general absence of cyclones below about latitude $25°$ for a time comparable to the lifetime ($\simeq 1000\,\text{yr}$) of an individual cell. Parker's model has been further developed by Levy (1972a, b, c), Cox (1975a) and Hoffman (1977, 1979). These will be considered in more detail later. Parker's model is not, like Rikitake's, a mechanism dependent on an exchange of energy between magnetic and

[†] Because the fluid in convective cells undergoes a twisting motion as it rises and falls due to Coriolis forces, the individual cells have been compared to cyclonic weather patterns.

mechanical forms—in fact, it has no explicit inertial component. It depends instead on the interaction of two almost independent self-sustaining dynamos magnetically coupled. They are assumed to vary individually in regenerating strength, either statistically or with secular change of physical parameters. One is idealised in the calculations as a pair of rings of upward streams, or cyclones, at high latitude near the pole in each hemisphere. The other is such a pair of rings at low latitude near the equator. Normally when they are both operating, or when the distribution of upwellings is random, they are parts of the same global dynamo. However, if one of them ceases to function, the direction of the magnetic field at its position is determined by the toroidal field generated by the other. This can locally have reversed direction and, when the dormant ring starts generating again, it will be excited in the reverse direction and may take over if the other ring weakens.

In both Cox's and Parker's models, cyclonic convection cells in the core produce reversals by a two-step mechanism. At any instant they are randomly distributed throughout the core; reversals occur when, through random processes, they arrive at certain configurations. However, there is a fundamental difference between the two models. In that of Parker the occurrence of a reversal depends only on the spatial distribution of the cyclones, and not on the intensity of the dipole field. In the model of Cox the occurrence of reversals depends upon both the distribution of the cyclones and on the field strength of the cyclone disturbances (i.e. the non-dipole field) relative to the dipole field.

Laj *et al* (1979) pointed out that there is a mechanical difficulty in Cox's model arising from the different time constants of the random fluctuations of the non-dipole field and the steady oscillations of the dipole field. Oscillations of the dipole field typically have a time constant of the order of 2×10^4 yr. Characteristic times for the non-dipole field range up to about 10^3 yr; however, most of the power is concentrated in the range 2–4×10^2 yr. It is quite difficult to couple two systems with very different time constants. Laj *et al* maintain that, if Cox's model is correct, the field would reverse but would maintain the opposite polarity for a very short time—of the order of the time constant of the non-dipole field. This problem of the differences between the time constants of the dipole and non-dipole fields also occurs in other models. In Parker's (1969) model the quantitative nature of the cyclones is not sufficiently well known to be able to calculate the minimum amount of time during which they have to be absent from low latitudes in order to induce a reversal. Parker believes that reversals occurring as often as 10^5 yr are not likely. McFadden and McElhinny (1982) showed that Cox's model gives a probability density function with a *minimum* around the central mean, whereas the palaeomagnetic data indicate a *maximum* around a central mean. Kono (1971, 1972) has also shown that Cox's model is not compatible with palaeointensity data for the last 10^7 yr and discussed various mathematical models which satisfy the statistical properties of the field variations during this time interval. In Cox's model the geomagnetic field is represented in the core by a large geocentric dipole and several off-centred dipoles; in Kono's final model, there are ten dipoles in the core with equal moments and directions either parallel or antiparallel to the rotation axis (see figure 5.10 for a schematic representation of Cox's and Kono's models). Inversion of each dipole takes place in a stochastic manner leading to variations in the polarity of the overall field. Kono stresses that his model may not be realistic physically, and may not be unique.

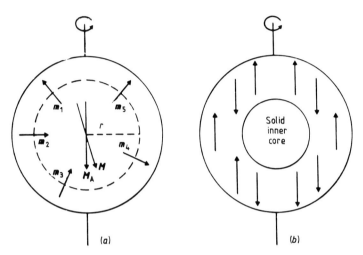

Figure 5.10 Schematic representation of (*a*) the model of Cox (1968) and (*b*) the mathematical model of Kono (1972). The arrows indicate dipoles distributed in the core (after Kono 1972).

Early palaeomagnetic work which sampled tens of rock strata indicated that the main dipole field was parallel to the Earth's rotation axis (although undergoing fluctuations in intensity of up to about $\pm 50\%$). When, later, sampling was extended to hundreds of rock strata at a given locality, the mean palaeofield was often found to differ slightly, but nevertheless significantly, from the field of an axial dipole (Cox and Doell 1964). The angular differences typically range up to about 6°, corresponding to a persistent component of the time averaged field of at least several thousand nanotesla. The observed average palaeofield is also rotated about 5° upward from the field of an axial dipole at localities within several tens of degrees of the equator. Detailed work on lava flows in Hawaii dating back to 5.6 Myr indicates that a non-dipole component of the time averaged field has existed for at least the past 5 Myr.

Cox (1975a) showed that this palaeomagnetic inclination anomaly is not due to a standing zonal field, but rather to the occasional drift beneath the sampling site of a giant magnetic anomaly similar to those which characterise most of the present non-dipole field. The palaeomagnetic data also indicate that in the past these anomalies were not generated at randomly distributed locations in the Earth's core. Most of the anomalies that occurred within about 30° of the equator appear to have been negative (in the sense of corresponding to the field of a dipole in the outer core directed radially outward). Most of the anomalies that occurred outside this zone were positive. The resulting pattern is one in which, during times of normal polarity, flux lines emerge from the core in localised regions near the equator and re-enter the core in localised regions within approximately 55° of the poles (see figure 5.11). The pattern is reminiscent of bipolar sunspot pairs, with the important difference that a spot from which magnetic flux emerges from the Sun is located on the same circle of heliographic latitude as the spot where flux re-

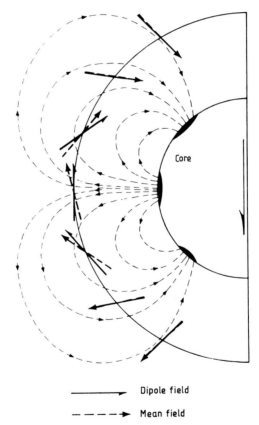

——————→ Dipole field

— — — —→ Mean field

Figure 5.11 Schematic representation of the non-dipole field (light broken curves) needed to account for the deflection of the time averaged palaeomagnetic field (heavy broken arrows) from the field of an axial dipole (heavy full arrows) (after Cox 1975a).

enters the Sun, whereas in the Earth spots of opposing polarity are at different latitudes. The long-term non-dipole field is antisymmetric in the sense that it also reverses direction when the main dipole field reverses. Thus, in Cox's model, during times when the field is reversed, magnetic flux emerges from the core in localised regions at high and mid latitudes and re-enters the core near the equator in perfect antisymmetry to the pattern that exists when the field is normal. If, as seems likely, the sign of the giant anomalies is related to the direction of flux in the toroidal field in the core, this suggests that when the dipole reverses, the polarity of the toroidal field also reverses.

A mechanism for reversals based on cyclonic convection in the core as proposed by Parker (1969) has been further developed by Levy (1972a, b, c). A steady dipole field is maintained by a two-stage process. Non-uniform rotation of the liquid core first draws the poloidal field into a toroidal magnetic field (see figure 5.12). The toroidal field is then twisted into meridional loops of poloidal field by cyclonic convective motions in the core. If a meridional loop has the same sense as the large-scale dipole, then the dipole field is regenerated. If a loop has the opposite sense, the dipole field is degraded. Levy showed that there are generally two types of toroidal zones in the core, regions of normal toroidal flux (where convection reinforces the main dipole), and regions of reverse toroidal flux (where convection degrades the poloidal field). Levy suggested two possible reversal schemes. In the first the dipole field is maintained by cyclonic convection concentrated at low latitudes. A fluctuation resulting in a large burst of cyclones in high latitudes suffuses the low latitude region with reverse toroidal flux. Normal regenerative low latitude

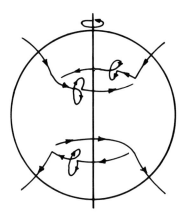

Figure 5.12 In the normal regeneration of the geomagnetic field, large scale velocity shear stretches the poloidal field into an azimuthal, toroidal field B_ϕ. Rising cyclones twist B_ϕ into poloidal loops of field. If these loops reinforce the original dipole field, then a regenerative dynamo can exist (after Levy 1972b).

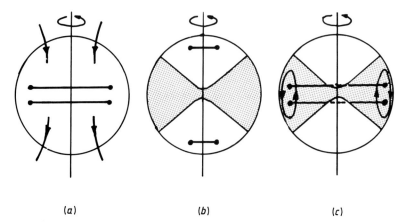

(a)   (b)   (c)

Figure 5.13   (*a*) A dipole field maintained by cyclones concentrated at low latitudes produces normal toroidal field throughout the core. (*b*) A fluctuation in the distribution of cyclones, consisting of a large burst of high latitude cyclones, produces a region of reverse toroidal flux at low latitude (shaded area). (*c*) Subsequent low latitude cyclones then generate poloidal field with sense opposite to the original dipole, thus reversing the dipole field (after Levy 1972b).

cyclones then create a dipole field with sign opposite to the original field (see figure 5.13).

In the second scheme the field is maintained normally by cyclones concentrated at high latitudes. In this case, the low latitude region of the core normally contains reverse toroidal flux. A fluctuation in which a large burst of cyclones appears at low latitudes directly produces poloidal field with sense opposite to the large-scale dipole field, thereby causing the dipole field to change sign (see figure 5.14). Although neither of these two schemes may actually occur, Levy maintains that a fluctuation in the distribution of cyclones with latitude can reverse the dipole field. Cox's (1975a) analysis of global palaeomagnetic data supports Levy's model in the sense that large non-dipole anomalies generally tend to maintain the same sign along a given circle of latitude. This observation is consistent with the existence of Levy's zones of reversed and normal toroidal flux. The symmetry of Levy's and Cox's models is different in the sense that in Cox's model the cyclonic cells at mid latitudes reinforce the dipole field in the northern hemisphere while those in the southern hemisphere degrade it, whereas in Levy's model reinforcement occurs at mid latitudes in both hemispheres. However, both models are antisymmetric with respect to polarity reversal in the sense that when the main dipole field reverses, the directions of the non-dipole fields and toroidal fields also reverse. In the Parker–Levy model no consideration is given to the question of how and why fluctuations in the distribution of cyclones occur.

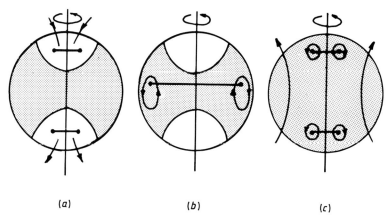

(a)   (b)   (c)

Figure 5.14 (*a*) A dipole maintained by cyclones concentrated at high latitudes produces both normal and reverse toroidal flux in the core. Shaded region represents reverse toroidal flux. (*b*) A burst of low latitude cyclones produces poloidal field with sense opposite to the original dipole. (*c*) It also floods the entire core with reverse toroidal flux so that subsequent high latitude cyclones then maintain the reversed field (after Levy 1972b).

Moreover even if some mechanism causes a field reversal, the field cannot grow again from a stray field when the steady state fluid motion is recovered—a steady state fluid motion can only maintain the existing magnetic field.

There have been a number of attempts at a numerical solution of the dynamo problem (e.g. Stevenson and Wolfson 1966, Kropachev 1971a, b). All these models include rotation and are drive by a radial body force. None of them have been successful because of the severe restriction on the size of the allowable timestep. Gubbins (1975) avoided this difficulty by neglecting rotation and assuming the fluid to have an appreciable viscosity. He found that if the dynamo was driven well above some critical limit, the magnetic field underwent large oscillations which might ultimately lead to a reversal. However the time required to complete a reversal was very long ($\sim 3 \times 10^4$ yr).

Watanabe (1981) has investigated the non-steady state of a simplified $\alpha\omega$ dynamo (see §1.4) allowing for Coriolis, Lorenz and viscous forces. He found that the dipole field undergoes relatively gradual variations with occasional reversals. The reversal of the dipole field is caused essentially by the disruption of the non-uniform rotation, the time required for a complete decay being about 3000 yr. The quadrupole field also changes sign and it appears that the toroidal field, from which poloidal fields (such as g_1^0, g_2^0, g_3^0) are produced by the effect, also reverses. Watanabe also found that field reversal is accompanied by an acceleration of the mantle and a change in the length of the day of about 95 ms.

5.4 The secular variation, reversals and polarity bias

In Cox's model there is a finite probability that the sum of the axial component of the non-dipole (ND) contributions will overcome the dipole (D) field causing a reversal. Hence the probability of reversal should be high if the ratio ND/D is large, and vice versa. If the dipole field is much larger than the sum of the axial components of the ND field, then the field is stable. As the ND fields increase, their axial components may overwhelm the dipole field, and the geomagnetic dynamo then amplifies the field in the opposite sense. Thus the magnitude of the palaeosecular variation should be related to the frequency with which reversals occur: when reversals are frequent the palaeosecular variation should be large and vice versa. Brock (1971) showed that such a relationship is generally borne out by the palaeomagnetic data. He also showed that the magnitude of the palaeosecular variation in pre-Cenozoic times was on the average about 15% less than during the Cenozoic and suggested that this difference is reflected in the reversal rate in the past. Irving and Pullaiah (1976) carried out a study of the palaeosecular variation in the later Phanerozoic (0–350 Myr) using a data base about twice as large as that available to Brock and confirmed his conclusions. They plotted overlapping polarity ratios for the Phanerozoic—see figure 5.15 which clearly illustrates polarity 'bias'. During long intervals of geologic time (as in the late Palaeozoic) it was reversed most of the time. Thus the assumption that the reversal process is symmetrical in the two polarity states with no bias with regard to the lengths of normal and reversed intervals is not justified. Irving and Pullaiah further showed that this cyclic change in polarity bias, noted previously by McElhinny (1971) is not due to the occurrence of quiet intervals superimposed on a general background of disturbed intervals in which the polarity is evenly balanced, because bias in the disturbed intervals themselves show the same cyclic change. The even balance between normal and reversed polarity during the Cenozoic reoccurs in the

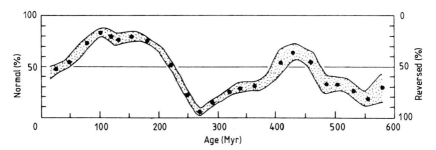

Figure 5.15 Polarity bias of the geomagnetic field during the Phanerozoic. Overlapping 50 Myr averages of polarity ratios as observed in palaeomagnetic results are shown together with the limits of the standard errors (after Irving and Pullaiah 1976).

Lower Triassic and in the Late Silurian to Lower Devonian, but over the entire Phanerozoic it is comparatively rare.

Crain *et al* (1969) computed a Fourier power spectrum of the proportion of normal polarity measurements using a compilation by Simpson (1966) with an effective sampling interval of about 15 Myr. They observed periodicities of 300 and 80 Myr. Later Crain and Crain (1970) carried out a spectral analysis of Simpson's mixed measurements and obtained periods of about 150 and 40 Myr. Fourier spectral analysis is, however, of limited usefulness in the treatment of short time series. Maximum entropy spectral analysis is a more powerful method, and Ulrych (1972) applied it to the analysis of polarity ratios. He took the data compiled by McElhinny (1971), fitted them to a ninth order polynomial, and interpolated with an effective sampling interval of 25 Myr. He obtained periodicities of 700 ± 100 Myr and 250 ± 50 Myr. Irving and Pullaiah (1976) also carried out a maximum entropy spectral analysis using overlapping averages of 5, 10, 25, 50 and 100 Myr. Figure 5.16 shows the normalised power spectra for the various sampling intervals—there are three major components with mean periodicities of 297 ± 34 Myr, 113 ± 5 Myr, and 57 ± 1 Myr. The latter two periodicities are not very evident in the spectrum from the 50 Myr averages, although there is some indication for them. Irving and Pullaiah were unable to reproduce the 700 Myr periodicity found by Ulrych (1972), and are sceptical about estimates of periods longer than the length of the record (600 Myr). The periodicity of about 300 Myr is common to the analyses of Crain *et al* (1969), Ulrych (1972), and Irving and Pullaiah (1976),

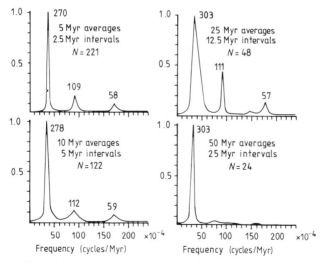

Figure 5.16 Maximum entropy power spectra for four averaging intervals. The ages of the peaks are given in Myr (after Irving and Pullaiah 1976).

and is evident visually in figure 5.15. Crain and Crain (1970) suggested that the periodicity around 300 Myr may be correlated with the Earth's position in the Galaxy, the rotational period of which is about 280 Myr. The shorter periodicities of 113 and 57 Myr are remarkably consistent, and presumably correspond to the variations superimposed on the general trend that can be seen in the 25 Myr averages. Creer (1975) has obtained a tentative correlation during Phanerozoic time between changes in polarity bias and changes in the length of the day determined by counts of growth rings on fossil shells.

It is interesting to speculate on the possible connection of the 300 Myr periodicity with the timescale of plate tectonics. The change from reversed to normal bias of the geomagnetic field coincident with the Palaeozoic–Mesozoic boundary is also one of the greatest breaks in the geologic record. Many suggestions have been made of a possible connection between changes in the Earth's magnetic field, whose origin lies in the core, and other geophysical phenomena occurring at the Earth's surface. Since only small variations in core motions are required to reverse the polarity of the dipole field, these could be caused by changes in the boundary conditions at the MCB (see also §3.5). Such boundary conditions could be affected by motions in the lower mantle and it is thus not implausible, as Hide (1967) has suggested, that reversals are correlated to some extent with other phenomena that may be affected by motions in the mantle. In this connection Hide and Malin (1970) found a correlation between the Earth's gravitational field and the ND field which they attributed to the effect of topography at the MCB. Irving (1966) has also suggested that the magnetic field would reverse frequently during times of active convection and tectonism (e.g. in the late Cenozoic). It is difficult to believe that every reversal is related to some event at the Earth's surface—the timescale of tectonic events is much longer than that for reversals. On the other hand it is not unreasonable to associate a marked change in the frequency of reversals (approximate timescale 50 Myr) with other phenomena.

Reid (1972) and Stewart and Irving (1973) found that reversal rates in the Precambrian varied in much the same way as those in the Phanerozoic. Reid reported a variation in the reversal rate from $0.4\,\mathrm{Myr}^{-1}$ to $1.1\,\mathrm{Myr}^{-1}$ over a 60 Myr interval, roughly 1800 Myr ago, and Stewart and Irving reversal frequencies less than $0.1\,\mathrm{Myr}^{-1}$ 990 Myr ago, and greater than $1\,\mathrm{Myr}^{-1}$ 790 Myr ago. Vogt *et al* (1972) reported 41 reversals between 150 and 135 Myr ago, giving an average reversal rate of $2.7\,\mathrm{Myr}^{-1}$ in the late Jurassic, and Helsley (1972) found that at least 23 reversals occurred during the Triassic, giving an average reversal rate greater than $0.7\,\mathrm{Myr}^{-1}$ between 225 and 190 Myr ago.

Jones (1977) has suggested that long-term variations in reversal frequency might be the result of fluctuating temperatures at the MCB caused by convection in the lower mantle. Numerical calculations at high Rayleigh numbers suggest that such convection may be intermittent (Foster 1971). In Jones' model a thermal boundary layer forms at the base of the mantle and grows by conduction. When it has reached a certain thickness, it becomes unstable and ejects hot material, thus destroying itself. Later a new boundary layer forms and the process is repeated. A thermal boundary layer at the base of the mantle could also account for the anomalous velocity gradients of P and S waves which are observed there. (The more generally accepted view (see e.g. Bolt 1972) has been that they are due to an increase in density arising from the incorporation of dense material from the core.) During the convective part of the intermittent cycle, temperatures along the MCB will show lateral variations caused by the pattern of ascending and descending currents, and also a decrease with time of the average temperature because of the destruction of the thermal boundary layer (Foster 1971). Jones argued, by analogy with the oceanic thermocline, that the effect of lateral temperature variations along the MCB would be confined to a thin boundary layer at the top of the core and would not cause any significant changes in reversal frequency. On the other hand a reduction in temperature at the MCB during the convective part of the cycle would increase the Rayleigh number of the core which could affect the flow pattern throughout a large part of the core. (Jones estimated a temperature gradient of $\sim 12 \, \text{K km}^{-1}$ in the thermal boundary layer.)

Cox (1981) has shown how stochastic models of the geomagnetic field can be generalised to take account of polarity bias. Defining polarity bias B as the fraction of time when the polarity is normal, during the Cenozoic $B \simeq 0.5$. However as already noted (figure 5.15), during the last part of the Mesozoic, approximately 110–80 Myr BP, the field had a strong normal bias ($B > 0.9$) while during the late Palaeozoic, approximately 310–230 Myr BP, the field had a strong reversed bias ($B < 0.1$). Intervals during which the bias remains constant are variable in length, ranging from 30 to more than 100 Myr. During times of strong polarity bias the frequency of reversals is generally low and during times of low bias ($B \simeq 0.5$) the frequency is high and sometimes variable, as in the Cenozoic. During times of extreme polarity bias, as in the late Mesozoic and late Palaeozoic, the dynamo is not locked into one polarity state. This is demonstrated by the presence in the late Mesozoic record of a few short, irregularly spaced reversed intervals and by the presence during the late Palaeozoic of a few short irregularly spaced normal intervals (Irving and Pullaiah 1976).

Cox (1981) generalised that class of stochastic models in which the field becomes unstable when, through some random processes, the magnetic field or fluid flow pattern in the core takes up some particular configuration (cf Cox 1968, Nagata 1969, Parker 1969). Cox assumed that Poisson distributed instabilities (see §5.5) occur at the same mean rate whatever the polarity state. He first constructed models which show bias with no memory, i.e. the field does not reverse its polarity prior to the onset of an instability; he later constructed models with both bias and memory, i.e. the probability of an instability not leading to a reversal depends upon the polarity state at the time of the instability, being different in the two polarity states. He finally constructed models in which Poisson distributed instabilities occur at different rates in the normal and reversed states. However, none of these models can explain such fundamental questions as why the geomagnetic field sometimes displays polarity bias, why the amount and sign of the bias change from time to time and why the frequency of reversals has changed throughout geologic time.

Cox (1981) suggested that symmetry is probably the key to understanding the question of bias. In the differential equations that describe the dynamo, there is no asymmetry with respect to the rotation axis. The total energy is also the same for both normal and reversed fields and both are equally probable. However, the non-dipole component of the field is highly irregular and asymmetric, and it is this field and the secular variation that Cox believes controls polarity bias. Since the required asymmetry is absent from the differential equations of the dynamo, it must reside in the boundary conditions at the MCB. An example of the type of required asymmetry in boundary conditions is any odd-zonal harmonic. A physical example would be zonal convection through the entire core produced by a difference in the temperature of the lower mantle in the northern and southern hemispheres. Cox further speculated that if lateral variations in temperature and density in the lower mantle partially control the pattern of fluid motions in the core, changes in reversal frequency may reflect shifts in the location of individual convective cells in the lower mantle, whereas changes in polarity bias may reflect changes in the entire pattern of mantle convection of the type that might, for example, be associated with a regrouping of the continents in one or the other hemisphere.

Finally it must not be forgotten that the observed polarity sequence is not necessarily the same as the polarity sequence of the core itself. This is because observations are not perfect and some reversals actually recorded in the Earth's rocks will have been missed. Again, the process by which rocks acquire their magnetisation acts as a filter to suppress the recording of short polarity intervals. Unfortunately the rock magnetic filtering process is not amenable to direct deconvolution and in the past there have been several attempts at indirect methods for determining the rate at which actual core reversals have occurred. Recently there has been some speculation that some of the instabilities in the dynamo process are what Cox (1981) has called 'infertile', i.e. they do not lead to reversals. There is some evidence for the existence of infertile instabilities during normal polarity (but not during reverse polarity) over the past 80 Myr; but during the period of 120 Myr back to 170 Myr most of the instabilities were fertile.

5.5 Statistical analyses

Variations in the length τ of polarity intervals may be described by a probability function derived from the theory of runs in Bernoulli trials (Cox

1968, 1969). If p is the probability of a polarity change during one cycle of the dipole field, then the probability that a polarity reversal will occur on the xth cycle after $(x-1)$ cycles of non-reversal is $p(1-p)^{x-1}$. Thus the probability function of τ is

$$P(\tau_p < \tau \leqslant \tau_p + \tau_D) = p(1-p)\tau_p/\tau_D \qquad (5.18)$$

where τ_D is the period of the dipole field, x is an integer, and $\tau_p = (x-1)\tau_D$. The mean value of τ for this distribution is

$$m = \tau_D/p. \qquad (5.19)$$

Cox (1968) used all available data for the past 10.6 Myr to plot the cumulative distribution (figure 5.17) together with curves of the theoretical cumulative distribution

$$P(\tau \leqslant \tau_c) = \sum_{i=1}^{n} p(1-p)^{i-1} \qquad (5.20)$$

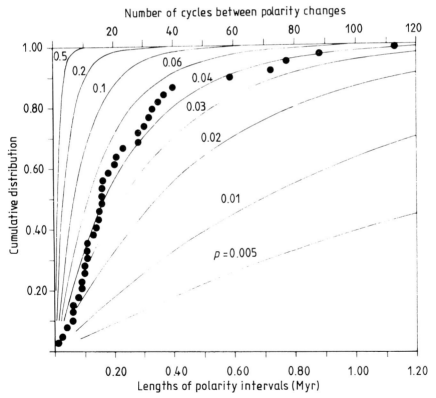

Figure 5.17 Cumulative distribution of polarity intervals. Circles are observational data. Curves are theoretical cumulative distributions obtained from equation (5.20) (after Cox 1968).

where n is an integer and $\tau_c = n\tau_D$. Almost all the data lie between the curves for $p = 0.03$ and 0.06. If the mean observed interval length of 0.22 Myr is taken as the theoretical mean m and τ_D as 10^4 yr, the corresponding value of p is 0.045.

If p is small and x is large in equation (5.18) $P(\tau)$ may be approximated by the Poisson distribution

$$P(\tau) = \lambda \exp(-\lambda\tau) \qquad (5.21)$$

where the parameter λ characterises the observed variations in the length of polarity intervals. The mean of this distribution is $m = 1/\lambda$ so that $\lambda = p/\tau_D$. For time intervals longer than some minimum time τ_{min}, equation (5.21) will be valid if and only if the probability of a reversal in any time interval $\tau > \tau_{min}$ is independent of whether a reversal occurred during any prior interval. Thus, if reversals are truly Poisson, the core has no memory longer than τ_{min}, which implies that the dynamo processes responsible for reversals have time constants no greater than τ_{min}.

τ_{min} is probably comparable to the longer time constants of the secular variation ($\sim 10^4$ yr). Cox (1969) also found that the apparent average duration of polarity intervals was greater during the time $10.6 < t < 45$ Myr BP than during the past 10.6 Myr, while during the time $45 < t < 75$ Myr BP the average length was longer still. It appears that the average length of polarity intervals (and hence the value of p) has changed during the Earth's evolution reflecting changes in the physical conditions in the Earth's core.

Nagata (1969) disagrees with Cox's suggestion that the main field can be represented by a steady dipole oscillator, and that a trigger effect takes place whenever the non-dipole field becomes sufficiently large relative to that of the dipole. He offered an alternative interpretation of the distribution function of reversals, based on the hypothesis that the main dipole field is steadily maintained only so long as the convection pattern in the core is asymmetric, as is the case in the Bullard–Gellman–Lilley dynamo (Lilley 1970a) but collapses when the convection pattern becomes symmetric, as in the original Bullard–Gellman model (1954). Nagata obtained a distribution function of the same mathematical form as that of Cox with $\lambda = \frac{1}{2}t_0$ where t_0 is the average lifetime for the polarity of a steady dipole. The average length of polarity intervals is $2t_0$. Lilley (1970b) suggested that the deviation of the geomagnetic axial dipole from true north is an expression of asymmetric motion in the Earth's core. A wandering of the magnetic poles to coincide with the geographic poles may thus be an indication of the flow in the core becoming symmetric. If this should happen dynamo action would be lost and the dipole field would decay, perhaps to grow again in the opposite direction (i.e. reverse) when the flow pattern once more becomes asymmetric. As a corollary, strong dipole fields may accompany a large deviation of magnetic north from true north. In this regard it is interesting to recall that cases have been reported when the geomagnetic field was large and stable when the VGP was close to the equator (see §3.2).

Similar statistical behaviour of the sequence of reversals is predicted by most models that have been proposed. In general, these models assume that reversals are due to instabilities in the MHD process and that an instability may take place at any time following a previous instability.

The exact physical nature of the instabilities varies with the model. They may result from too much symmetry in the fluid motions of the core (Nagata 1969) or from some critical configuration of convection cells (Parker 1969, Levy 1972b, c). The models presume that the core retains no memory of past fluid configurations beyond the average lifetime of the MHD process. This condition defines a renewal process. Because the average time intervals between successive reversals are much longer than the time constants of the dynamo, further simplification can be made by assuming that the probability of a reversal per unit time is a constant independent of the time of the previous reversal. In this case, reversals describe a Poisson process (Cox 1968, 1969, 1970, Nagata 1969), and the time intervals between successive reversals show an exponential distribution (see equation (5.21)). Although reversal timescales obtained from marine magnetic anomalies do show some properties of renewal processes (Phillips *et al* 1975, Phillips and Cox 1976), they cannot be used in a direct test of the Poisson model. This is because the effects of short polarity events, which may account for nearly half the true number of reversals, are difficult to detect in marine magnetic anomalies. It is likely that a large number of these short events are missing from even the most recently determined timescales. This question is discussed again later.

The observed sequence of reversals for the past 70 Myr satisfies a Poisson distribution reasonably well except that the observed number of short polarity intervals is somewhat smaller than predicted (Cox 1969, Naidu 1971). With the discovery of previously undetected short events (Blakely and Cox 1972) the fit between the observed and predicted distributions has improved but still is not perfect. It is uncertain whether this is because of the failure to detect a significant number of short events in the palaeomagnetic record or because the underlying geomagnetic reversal process is not perfectly Poisson (Harrison 1969, Cox 1969, Aldridge and Jacobs 1974, Tacier *et al* 1975)[†]. Even if reversals turn out to be non-Poisson, they may still be generated by any of several other classes of stochastic processes. For a Poisson process $p(t)$ is

† Aldridge and Jacobs (1974) constructed mortality curves for each normal and reversed phase of the Earth's magnetic field over the past 45 Myr which revealed a departure from the simple exponential distribution for the lifetime of these phases. The significance of this departure was demonstrated by comparing simulated distributions of an exponential random variable with actual distributions of the normal and reversed phases. Selective removal of shorter phases from the sequence of exponentially distributed random variables distorted the mortality curves for these generated phases so that they resembled the mortality curves of the observed phases. This demonstration provides evidence that there are probably undetected short phases throughout the past 45 Myr.

constant, whereas for a general renewal process $p(t)$ varies so that the probability of a reversal increases or decreases with increasing time t; however $p(t)$ depends *only* on the time since the previous reversal and not on the length of prior polarity intervals. Thus if geomagnetic reversals are generated by a renewal process, the lengths of polarity intervals will be independent. If normal and reversed polarity intervals have different mean lengths, the lengths of polarity intervals will not be independent. This question will be discussed later.

Naidu (1971) suggested that the probability density function for reversals is better described by a gamma distribution

$$P(\tau) = \lambda(\lambda\tau)^{k-1} \exp(-\lambda\tau)/\Gamma(k) \tag{5.22}$$

for which the mean value $\mu = k/\lambda$. This reduces to equation (5.21) if the lengths of polarity intervals are Poisson distributed ($k = 1$). The parameter k measures both the proximity of the observed distribution to an exponential distribution and the dispersion of the intervals about their mean. Naidu's original purpose (1974) was to see whether polarity intervals were independent. He came to the conclusion that for the period 0–48 Myr BP they were not independent and that harmonic components with a fundamental frequency of 0.75 cycles/Myr were present in the data. He offered no physical explanation for this. He extended his work in a later paper (1975) and claimed that reversal intervals for the period 0–76 Myr BP were not independent—implying that the geomagnetic dynamo possesses a memory. He developed his model on the basis of the exponential nature of the autocorrelation function for the entire 76 Myr data set. The main reason for the difference in Naidu's (1975) and Phillips *et al*'s (1975) conclusions (see §3.5 and figure 3.22) regarding the independence of polarity intervals is that Naidu analysed the entire data set from 0–76 Myr whereas Phillips *et al* subdivided the data set into two subsets in each of which the field was stationary. They found that the mean interval length is 0.33 ± 0.03 Myr during the last 45 Myr and 0.90 ± 0.12 Myr during the early Cenozoic. Ulrych and Clayton (1976) pointed out that implicit in Naidu's model is the assumption of stationarity during this time interval. Naidu (1971) himself showed that the statistical structure of the polarity intervals underwent a marked transition about 48 Myr ago—this has a pronounced effect on the computed autocorrelation function. Ulrych and Clayton clearly demonstrated the independence of polarity intervals over this timespan—the dependence found by Naidu being a result of the discontinuity in the statistical properties of the data around 48 Myr BP. Naidu later (1976) agreed with them but still believes that for the period 48–72 Myr BP, reversal intervals are not independent. Phillips and Cox (1976) have also examined the spectrum of reversal timescales, obtaining a generalised expression for the power density spectrum of a random telegraph wave with gamma distributed polarity intervals. They were unable to reproduce Naidu's expression for the theoretical spectrum nor could they find any periodicity.

If the Earth's field is not generated by a Poisson process, the probability of a

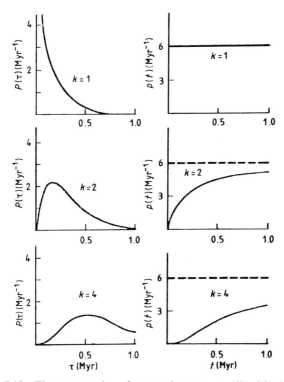

Figure 5.18 Three examples of renewal processes all with the same average frequency of reversals of $6 \times 10^{-6} \, \mathrm{yr}^{-1}$. Curves on the left are probability density functions for the lengths of polarity intervals. Curves on the right show the instantaneous probability that a reversal will occur per unit time. The upper two curves describe a Poisson renewal process ($k=1$). The lower four curves describe renewal processes in which the instantaneous probability that a reversal will occur increases with the passage of time after a reversal (after Cox 1975b).

reversal per unit time drops to zero immediately after a reversal and subsequently rises slowly to a steady state value. Figure 5.18 shows three theoretical curves, all with the same mean frequency of reversals ($6 \, \mathrm{Myr}^{-1}$) but with varying values of k. For a Poisson reversal process ($k=1$), the instantaneous probability of a reversal per unit time rises immediately to the steady state value of $6 \times 10^{-6} \, \mathrm{yr}^{-1}$. For $k=2$, the 50% rise time is about $2 \times 10^5 \, \mathrm{yr}$ and for $k=4$, it is about $8 \times 10^5 \, \mathrm{yr}$. The existence of a non-zero rise time should not be too surprising. The Earth's magnetic field would not be driven to reverse unless the new polarity represented a temporarily lower and thus more stable energy state. Because the core fluid does retain some coherency over periods of at least $10^3 \, \mathrm{yr}$, we should expect this increased stability to last a while.

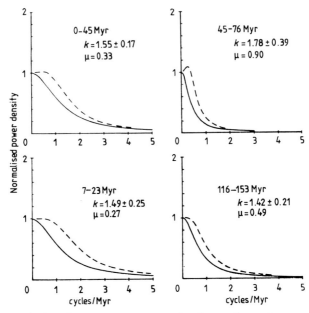

Figure 5.19 Maximum entropy spectral estimates for various geomagnetic reversal timescales (full curves). The Cenozoic timescale of Heirtzler *et al* (1968) has been divided at 45 Myr into two stationary segments (0–45 Myr and 45–76 Myr). Also shown are the spectral estimates obtained from the Miocene (7–23 Myr) reversal timescale of Blakely (1974) and the non-stationary Keathley reversal sequence (116–153 Myr) of Larson's Mesozoic reversal timescale (Larson 1974). Maximum likelihood estimates of k, its standard error, and the mean interval length μ are indicated in each case. Theoretical spectra corresponding to these values are shown in each case (broken curves) (after Phillips and Cox 1976).

Phillips and Cox (1976) have carried out maximum entropy spectral estimates for various reversal timescales (see figure 5.19—theoretical spectra corresponding to sample values of μ and k are also shown). The Cenozoic reversal timescale of Heirtzler *et al* (1968) has been divided into two stationary segments, 0–45 Myr and 45–76 Myr. The spectral estimates for the two segments reflect the change in the mean interval length that occurred around 45 Myr ago. Also shown is the spectral estimate for the Miocene (7–23 Myr) reversal timescale of Blakely (1974). Although the sample mean and k are smaller for this more complete data set, the spectrum appears identical to that obtained for the longer Cenozoic record. A final spectral estimate is presented for the Keathley reversal sequence of Larson's (1974) Mesozoic timescale. Although this sequence displays non-stationary behaviour, the low frequency variation of the mean interval length does not appear to affect the spectrum. All the spectra shown in figure 5.19 resemble the theoretical spectrum of a

Poisson reversal process. From studies on synthetic data, Phillips and Cox found that the behaviour is consistent with a gamma process having a k value between one and two. This is also true for the Mesozoic data in which the mean interval length is known to undergo a low frequency variation.

In a later paper, Phillips (1977) found that the parameter k undergoes a slow variation between 1.2 and 2.6. There is no evidence for discontinuities in k—the average value of k for the entire past 76 Myr is 1.72 ± 0.17. The fact that k does not seem to be affected by the change in mean interval length indicates that μ and k are independent. Phillips carried out a similar analysis for the Keathley reversal sequence of Larson's (1974) Mesozoic timescale (116–153 Myr) which is bounded on both sides by long intervals of normal polarity. During this time μ shows a slow variation (between 0.32 and 0.66 Myr, average 0.49 ± 0.05 Myr), with a period of about 25 Myr, which is inconsistent with stationarity. Unlike μ, the parameter k remains approximately constant with an average value of 1.64 ± 0.24 which is very close to the average Cenozoic value. The statistics of the Cenozoic and Mesozoic reversal timescales thus show that the mean interval length μ undergoes both sudden and gradual changes and has stable regions lasting for tens of Myr. Presumably μ is related to the growth and decay of features at the MCB which must therefore have similar time constants. The observed reversal process has a nearly constant value of k. Thus it appears that the current reversal process, which has operated throughout the Cenozoic, is not significantly different from that which operated during the Early Cretaceous and Late Jurassic. The Late Cretaceous long period of normal polarity separating these reversal sequences represents a very different state, which may also be related to special conditions at the MCB.

If the reversal sequences during the Cenozoic and middle Mesozoic can be modelled with a gamma reversal process having a time varying μ and a constant k, separate analyses of the reversed and normal polarity intervals should lead to the same conclusions. Phillips (1977) examined the Cenozoic reversal sequence for this purpose and found that the means of normal and reversed intervals behaved in a similar way; both show the discontinuity at 45 Myr, and both are consistent with stationary behaviour on either side of this discontinuity. Statistical tests established that there is no significant difference between the mean lengths of normal and reversed intervals during the Cenozoic. For k he found that for intervals of a particular polarity the sample values were all close to the long-term averages. However the average value of k for normal intervals is 2.84 ± 0.41, whereas for reversed intervals it is 1.31 ± 0.18. The most surprising feature of his analysis is this difference in the values of k for normal and reversed intervals—the values for normal intervals being consistently larger than those for reversed intervals. This difference implies that normal and reversed polarity intervals have different distributions. If this is the case reversals cannot be described as a simple renewal process but require a more complicated alternating renewal model. A further analysis of the Miocene (8–23 Myr) reversal timescale of Blakely (1974), the most complete long reversal sequence available, confirms these earlier findings—the mean lengths of normal and reversed intervals are virtually the same and the value of k determined for normal polarity intervals is greater than that determined for reversed polarity intervals. This persistent difference in k for normal and reversed intervals is evidence that the geodynamo is more stable after a transition to normal polarity than it is after a transition to reversed polarity. There is no evidence to suggest that this difference results from the physical recording process in the oceanic crust, so that it appears that the asymmetry is a property of the geomagnetic dynamo. The question of differences between normal and reversed palaeomagnetic fields has also been discussed in §3.2.

Lowrie and Kent (1983) disagree with the analyses of Cox (1975a) and Phillips (1977) which indicated a fairly abrupt increase in the frequency of reversals around 45 Myr ago[†]. Instead they found that the mean reversal frequency in the Cenozoic and Late Cretaceous is characterised by a fluctuation, with a period of about 20 Myr, superimposed on an almost linearly increasing trend. They thus claim that subdivision into two periods of stationary behaviour is no longer warranted. They based their conclusions on the newer magnetic polarity timescales of LaBrecque *et al* (1977), Ness *et al* (1980) and Lowrie and Alvarez (1981)—the analyses of Cox (1975a) and Phillips (1977) used the older magnetic timescale of Heirtzler *et al* (1968). A major difference arises from the omission of anomaly 14 seen by Heirtzler *et al*, but not found on most marine magnetic profiles. Anomaly 14 represents four very short normal polarity zones and thus introduces eight additional reversal boundaries. Lowrie and Kent showed that the 0–40 Myr polarity intervals may be represented by a gamma distribution, with index $k=2$. If this is substantiated, it would imply that the probability of a reversal per unit time is not constant as assumed by Cox (1968, 1969). They further speculated that there may be secular changes in core conditions during the past 80 Myr and that core processes may have a long-term memory for polarity reversals.

Lowrie and Kent also discussed the effect of adding short events to the magnetic polarity timescale—this has a dramatic effect on the polarity interval distribution between 0–40 Myr. The number of polarity intervals increases from 126 to 226 and the mean length of a polarity interval is reduced from 0.314 Myr to 0.175 Myr. The exact duration of such short polarity events cannot be determined. If they are 20 000 yr long, the polarity interval length distribution differs significantly from a Poisson distribution—if they are about 30 000 yr long, there is no significant difference. The effect of additional short polarity events also dramatically affects the distribution of normal and reversed polarity interval lengths. If the short events are about 30 000 yr long, the gamma index k for normal and reversed distributions is 1.14 and 2.01 respectively. Although different k values for normal and reversed distributions

† See also criticism by McFadden (1984) discussed in §3.5.

had been found before (e.g. Phillips 1977), the addition of these short polarity events alters the sense of asymmetry between normal and reversed states— previous analysis had suggested k was larger for normal polarity. This change results from the unequal distribution of the short polarity events which are predominantly of positive polarity and concentrated in the Late Cenozoic[†].

The origin of such short-wavelength magnetic anomalies is vital for statistically modelling magnetic polarity reversals. It is not known how many of the 57 possible small events indicated by LaBrecque *et al* (1977) are coherent. They have been interpreted as real polarity changes (Blakely and Cox 1972, Blakely 1974) or as magnetic field intensity fluctuations (Cande and LaBrecque 1974). More recently Schouten and Denham (1979) suggested that they may result from variability in the magnetised layer in high spreading rate areas. Lowrie and Kent admit that it is not possible to decide whether these short-wavelength features represent real polarity changes, intensity fluctuations or source layer variations.

Most mathematical descriptions of convection-driven dynamos are not asymmetric with respect to polarity. These models assume a dynamo that is spherical and axisymmetric. Kinematic solutions to the dynamo equations commonly assume symmetry about the equatorial plane as well. Slight departures from this geometry might conceivably affect the symmetry of normal and reversed states, but it is difficult to see how such departures could be maintained over the 150 Myr period of observations. All evidence suggests that reversals can be described by an alternating renewal process having a time-varying mean and gamma distributed polarity intervals. One of the difficulties in a statistical study of reversals is the extreme sensitivity of a histogram of polarity changes to the number of short polarity events. This sensitivity is due to the fact that the Earth's magnetic field has only two polarity states; thus a short polarity event inserted in the middle of a long period of opposite polarity not only adds the short polarity event to the histogram but also erases a long period and adds two middle-length periods. Laj *et al* (1979) analysed the series of geomagnetic reversals using a correlation function of the telegraph signal obtained from the Heirtzler timescale by assigning ± 1 to intervals of normal/reversed polarity. Unlike the function used by Naidu (1971, 1975), their autocorrelation function is only very slightly sensitive to undetected short polarity intervals. Thus their results are valid, irrespective of failure to detect short polarity intervals, and give information about the dynamics of the reversal time series. Laj *et al* also showed that other models, which do not need any triggering mechanism as in Cox's model, are compatible with the observed results. This is important, since, as already noted, it is very difficult to trigger a long period oscillator using short-period stochastic fluctuations of reasonable amplitude. Laj *et al* concluded that, within statistical noise limits, successive polarity intervals are independent and

† McFadden and Merrill (1984) have since examined the whole question of the value of k for normal and reversed states and concluded that there is no difference.

distributed in time according to a Poisson process, confirming the earlier results of Cox and his co-workers.

McFadden and McElhinny (1982) have analysed all VDM determined for the past 5 Myr which have associated VGP with latitudes greater than 45° (to ensure that they relate to periods of stable polarity). Histograms of normal, reversed and combined polarity data show that, although the data show maximum frequency around some central mean value, there is a fairly sharp cut-off in the observed VDM at the low VDM end and the data are skewed to the high end. McFadden and McElhinny showed that neither a Gaussian nor a lognormal distribution provides a satisfactory fit to the data. Their preferred model is one in which the TDM (true dipole moments) have a truncated Gaussian distribution (which they call nested) and the field strength of the non-dipole components is linearly proportional to the TDM (see figure 5.20). They

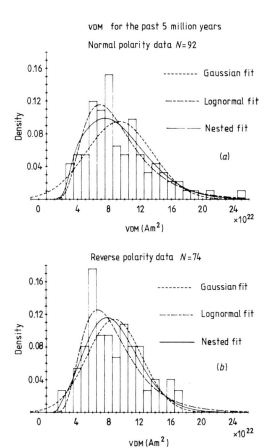

Figure 5.20 Nested distribution fit to VDM data with Gaussian and lognormal fits for comparison: (*a*) normal polarity, (*b*) reversed polarity (after McFadden and McElhinny 1982).

found no support for a model in which the field strength of the non-dipole component is a constant ratio of the mean dipole field strength, nor for Cox's (1968) model of a cyclic variation of the dipole moment during stable polarity periods.

McFadden and McElhinny further analysed the data to see whether there were any differences between the statistical parameters of the normal and reversed polarity distributions. They showed that the data give no reason to reject the hypothesis of a common mean and variance for the (untruncated) Gaussian distribution of the normal and reversed polarity TDM, although a common truncation point can be rejected—the truncation point for normal polarity TDM being larger (see figure 5.21). This indicates that there is a difference between the properties of the two polarity states. It could be interpreted as meaning that a stable polarity can be sustained for lower dipole moments in the reversed polarity state than in the normal polarity state, implying that the reversed state is inherently more stable than the normal state. Alternatively it could be interpreted as meaning that in the normal polarity state a stable polarity can be sustained for larger deviations of the VGP from the spin axis than in the reversed polarity state, implying that the normal state is inherently more stable than the reversed state.

5.6 Non-dipole components of the palaeomagnetic field

Although the present geomagnetic field deviates substantially from that of a geocentric axial dipole, it is usually assumed in palaeomagnetic studies that

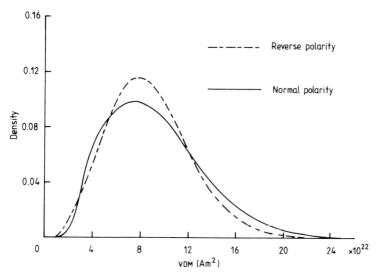

Figure 5.21 Comparison of nested distribution fits to normal and reversed polarity VDM data (after McFadden and McElhinny 1982).

such deviations are negligible when averaged over several thousand years. However, McElhinny and Merrill (1975) noted that nearly half of the palaeomagnetic poles derived from rocks during the last 5 Myr have 95% circles of confidence about them that do not include the present geographic poles. Wilson and Ade-Hall (1970) first noted that late Tertiary and Quaternary palaeomagnetic poles from Europe and Asia all tended to lie too far away from the observation site along the great circle joining the site to the geographic pole. Successive analyses of worldwide data by Wilson (1970, 1971, 1972) and Wilson and McElhinny (1974) confirmed that this far-sided effect occurs globally. It can be produced by an axial dipole that is not geocentric but is displaced a distance d northward along the axis of rotation. For the time-averaged late Tertiary field, Wilson and McElhinny obtained a value for d of 325 ± 57 km. As shown in §1.2, if the potential due to an offset axial dipole is expanded in terms of spherical harmonics only zonal terms occur. For $d = 325$ km, g_3^0 is less than 8% of the magnitude of g_2^0, so that the offset dipole is approximately equivalent to a spherical harmonic model including only the g_1^0 and g_2^0 terms.

Using different techniques and different data sets, Creer *et al* (1973), Wells (1973) and Merrill and McElhinny (1977) all confirmed the existence of a significant g_2^0 component in the Quaternary and Late Tertiary palaeofield. Hence the existence of at least one long-term non-dipole component appears to be well established. However, general agreement on the significance of any other long-term non-dipole components is lacking. Deviations from a geocentric axial dipole field can be expressed only in terms of ratios of Gauss coefficients because of the lack of corresponding palaeointensity data (see table 5.1). Merrill and McElhinny (1977) were unable to give an accurate estimate of the errors involved in estimating the coefficients in table 5.1: however, they were able to show that the north–south asymmetry, as reflected by the g_3^0/g_1^0 ratio, is significant at the 90% confidence level.

They suggested that this north–south asymmetry may reflect differences in

Table 5.1 Gaussian coefficients normalised to the geocentric axial dipole term $g_1^{0\dagger}$.

Term	Present field	Palaeomagnetic field	
		Normal	Reversed
g_1^1/g_1^0	$+0.067$	-0.017	$-\,?$
h_1^1/g_1^0	-0.190	$+0.030$	$+\,?$
g_2^0/g_1^0	$+0.063$	$+0.050$	$+0.083$
g_3^0/g_1^0	-0.043	-0.017	-0.034

† The term g_1^0 is negative for the present (normal) field and positive for the reversed field. The table therefore indicates that the sign of all the coefficients changes with that of g_1^0.

boundary conditions at the MCB such as temperature, topography and electrical conductivity. It has also been suggested that boundary conditions at the MCB may change with time (Cox 1968, McElhinny 1971, Phillips and Cox 1976). This idea has been reinforced by theoretical calculations by Jones (1977) involving a thermal boundary layer at the bottom of the mantle (see §5.4), indicating that times of the order of 50 Myr can be expected for changes to take place in boundary conditions at the MCB. Table 5.1 also shows that the magnetic field during times of normal polarity is significantly different (at the 95% confidence level) from that during times of reversed polarity.

Table 5.2 (after Coupland and Van der Voo 1980) gives the results of a spherical harmonic analysis of the Earth's magnetic field for a number of time periods using all available data for the last 130 Myr (meeting certain criteria), returned to precontinental drift site locations and orientations using spreading poles based on sea-floor magnetic anomalies. Coupland and Van der Voo showed that the discontinuity in reversal frequency at 45 Myr correlates almost exactly with a sharp drop in the magnitude of the non-dipole field at that time. Not only did the magnitude of the non-dipole field change, but its configuration changed as well. From about 95 to 50 Myr ago, the long-term non-dipole field consisted of a large g_2^0 component, while g_3^0 was near zero. Between 50 and 40 Myr ago, $g_2^0/|g_1^0|$ apparently changed from about -0.10 to $+0.05$, while the $g_3^0/|g_1^0|$ component changed from near zero to -0.10. Changes since then have been much more gradual. The large values of g_3^0 are incompatible with an offset dipole as a physical model for the generation of the long-term non-dipole field.

Table 5.2 Results of spherical harmonic analyses of the Earth's magnetic field.

Time period	Extent (Myr)	Data set	No of sites	g_2^0	g_3^0
Upper Cenozoic	0–26	extended	251	−0.0549	−0.0381
		restricted	102	−0.0488	−0.0426
Quaternary	0–2	extended	81	−0.0633	−0.0469
		restricted	45	−0.0581	−0.0522
Pliocene	2–7	extended	67	−0.0916	−0.0201
		restricted	18	−0.0767	−0.0661
Miocene	7–26	extended	58	0.0034	−0.0793
		restricted	23	−0.0081	−0.0463
Lower Tertiary	26–65	extended	51	−0.1343	
		restricted	22	−0.1552	
Upper Cretaceous	65–100	extended	42	−0.1279	
		restricted	18	−0.1446	
Lower Cretaceous	100–136	extended	34	(0.1952)	
		restricted	11	(−0.0213)	

Between 100 and 75 Myr ago, $g_2^0/|g_1^0|$ apparently increased from near zero to about -0.20. Thus there is some indication of a change in the configuration of the non-dipole field which correlates with the 85 Myr discontinuity in reversal frequency. However, the configuration of the non-dipole field is much less accurately determined in this time region than near the 45 Myr discontinuity, because of the relative paucity of palaeomagnetic results for rocks of Cretaceous age. Coupland and Van der Voo also found that over the period 10–50 Myr BP the value of g_2^0 determined for dominantly reversed data is about twice as large (in magnitude) as it is for dominantly normal data. This is in agreement with the results of Wilson (1972), who found that the dipole offset

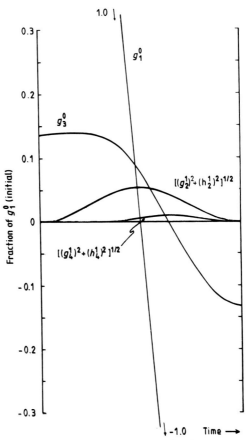

Figure 5.22 Variation in the three most dominant non-dipole field components present in the modelled transition field for the Brunhes–Matuyama transition; each term is normalised to the full polarity strength of the axial dipole. The decay through zero of the axial dipole is also indicated (after Hoffman 1981).

calculated for 24 reversed Quaternary and Upper Tertiary directions from the USSR was much greater than the dipole offset calculated for 36 normal directions (1050 km for the reversed field, 175 km for the normal field). However, if g_3^0 is included in the analysis, the magnitude of g_2^0 is no longer greater in the dominantly reversed data, nor is g_3^0 greater.

Hoffman (1979) simulated the Brunhes–Matuyama transition with a model based on the assumption that reversals start in a localised region of the core which extends or 'floods', both north–south and east–west until the entire core is affected (see §3.4). Assuming the reversal process to start at the equator, he later (1981) analysed his model solution to determine the behaviour of the dominant Gauss coefficients during the transition. Figure 5.22 shows the relative strengths of the three most dominant non-dipole terms (each normalised to the full polarity strength of the axial dipole) plotted versus time during the reversal. As the axial dipole decreases through zero these controlling terms are a zonal octupole g_3^0, a non-axisymmetric quadrupole having strength $[(g_2^1)^2 + (h_2^1)^2]^{1/2}$ and, to a far lesser extent, a higher order term having strength $[(g_4^1)^2 + (h_4^1)^2]^{1/2}$. Although not unique, Hoffman's estimate of the most significant Gauss components and their variation for the modelled transition field corresponding to the Brunhes–Matuyama reversal, is not tied to the phenomenological model from which they were derived, e.g. the strength of the non-dipole components at the time of complete dipole decay could result from a standing (i.e. non-reversing) field.

References

Aldridge K D and Jacobs J A 1974 Mortality curves for normal and reversed polarity intervals of the Earth's magnetic field *J. Geophys. Res.* **79** 4944

Allan D W 1958 Reversals of the Earth's magnetic field *Nature* **182** 469

——1962 On the behaviour of systems of coupled dynamos *Proc. Camb. Phil. Soc.* **58** 671

Blakely R J 1974 Geomagnetic reversals and crustal spreading rates during the Miocene *J. Geophys. Res.* **79** 2979

Blakely R J and Cox A 1972 Evidence for short geomagnetic polarity intervals in the early Cenozoic *J. Geophys. Res.* **77** 7065

Bochev A 1969 Two and three dipoles approximating the Earth's main magnetic field *Pure Appl. Geophys.* **74** 29

Bolt B A 1972 The density distribution near the base of the mantle and near the Earth's centre *Phys. Earth Planet. Int.* **5** 301

Brock A 1971 An experimental study of palaeosecular variation *Geophys. J.* **24** 303

Bullard E C 1955 The stability of a homopolar dynamo *Proc. Camb. Phil. Soc.* **51** 744

——1978 The disk dynamo *Topics in non-linear dynamics* ed S Jorna *AIP Conf. Proc.* No 46

Bullard E C and Gellman H 1954 Homogeneous dynamos and terrestrial magnetism *Phil. Trans. R. Soc.* A **247** 213

Cande S C and LaBrecque J L 1974 Behaviour of the Earth's palaeomagnetic field from small scale marine magnetic anomalies *Nature* **247** 26

Chandrasekhar S 1961 *Hydrodynamic and hydromagnetic stability* (Oxford: Clarendon) Ch V

Cook A E and Roberts P H 1970 The Rikitake two disc dynamo system *Proc. Camb. Phil. Soc.* **68** 547

Coupland D H and Van der Voo R B 1980 Long-term non-dipole components in the geomagnetic field during the last 130 Myr *J. Geophys. Res.* **85** 3529

Cox A 1968 Length of geomagnetic polarity intervals *J. Geophys. Res.* **73** 3247

——1969 Geomagnetic reversals *Science* **163** 237

——1970 Reconciliation of statistical models for reversals *J. Geophys. Res.* **75** 7501

——1975a The frequency of geomagnetic reversals and the symmetry of the non-dipole field *Rev. Geophys. Space Phys.* **13** 35

——1975b Symmetric and asymmetric geomagnetic reversals as a renewal process *Proc. Takesi Nagata Conference: Magnetic fields, past and present* (Pittsburgh: Univ. Pittsburgh)

——1981 A stochastic approach towards understanding the frequency and polarity bias of geomagnetic reversals *Phys. Earth Planet. Int.* **24** 178

Cox A and Doell R R 1964 Long period variation of the geomagnetic field *Bull. Seism. Soc. Am.* **54** 2243

Crain I K and Crain P L 1970 New stochastic model for geomagnetic reversals *Nature* **228** 39

Crain I K, Crain P L and Plaut M G 1969 Long period Fourier spectrum of geomagnetic reversals *Nature* **223** 283

Creer K M 1975 On a tentative correlation between changes in the geomagnetic polarity bias and reversal frequency and the Earth's rotation through Phanerozoic time *Growth Rhythms and the History of the Earth's Rotation* ed G D Rosenberg and S K Runcorn (New York: Wiley)

Creer K M, Georgi D T and Lowrie W 1973 On the representation of the Quaternary and Late Tertiary geomagnetic fields in terms of dipoles and quadrupoles *Geophys. J.* **33** 323

Creer K M and Ispir Y 1970 An interpretation of the behaviour of the geomagnetic field during polarity transitions *Phys. Earth Planet. Int.* **2** 283

Dagley P and Lawley E 1974 Palaeomagnetic evidence for the transitional behaviour of the geomagnetic field *Geophys. J.* **36** 577

Foster T D 1971 Intermittent convection *Geophys. Fluid Dyn.* **2** 201

Gubbins D 1975 Numerical solution of the hydromagnetic dynamo problem *Geophys. J.* **42** 295

Harrison C G A 1969 What is the true rate of reversals of the Earth's magnetic field? *Earth Planet. Sci. Lett.* **6** 186

Heirtzler J R, Dickson G O, Herron E M, Pitman W C and Le Pichon X 1968 Marine magnetic anomalies, geomagnetic field reversals and motions of the ocean floor and continents *J. Geophys. Res.* **73** 2119

Helsley C E 1972 Post-Palaeozoic magnetic reversals *Trans. Am. Geophys. Union* **53** 363

Herzenberg A 1958 Geomagnetic dynamos *Phil. Trans. R. Soc.* A **250** 543

Hide R 1967 Motions of the Earth's core and mantle, and variations in the main geomagnetic field *Science* **177** 55

Hide R and Malin S R C 1970 Novel correlations between global features of the Earth's gravitational and magnetic fields *Nature* **225** 605

Hoffman K A 1977 Polarity transition records and the geomagnetic dynamo *Science* **196** 1329

——1979 Behaviour of the geodynamo during reversal: a phenomenological model *Earth Planet. Sci. Lett.* **44** 7

——1981 Quantitative description of the geomagnetic field during the Matuyama–Brunhes polarity transition *Phys. Earth Planet. Int.* **24** 229

Huppert H E and Moore D R 1976 Non-linear double diffusive convection *J. Fluid Mech.* **78** 821

Irving E 1966 Palaeomagnetism of some Carboniferous rocks of New South Wales and its relation to geological events *J. Geophys. Res.* **71** 6025

Irving E and Pullaiah G 1976 Reversals of the geomagnetic field, magnetostratigraphy and relative magnitude of palaeosecular variation in the Phanerozoic *Earth Sci. Rev.* **12** 35

Ito K 1980 Chaos in the Rikitake two-disc dynamo system *Earth Planet. Sci. Lett.* **51** 451

Jones G M 1977 Thermal interaction of the core and the mantle and long-term behaviour of the geomagnetic field *J. Geophys. Res.* **82** 1703

Kerridge D J and Wilkinson I 1983 Spectral analyses of the reversal waveforms of the magnetic field produced by an experimental homogeneous dynamo *Geophys. J.* **72** 310

Kono M 1971 Intensity of the Earth's magnetic field during the Pliocene and Pleistocene in relation to the amplitude of mid-ocean ridge magnetic anomalies *Earth Planet. Sci. Lett.* **11** 10

——1972 Mathematical models of the Earth's magnetic field *Phys. Earth Planet. Int.* **5** 140

Kristjansson L and McDougal I 1982 Some aspects of the late Tertiary geomagnetic field in Iceland *Geophys. J.* **68** 273

Kropachev E P 1971a Numerical solutions of the equations of the dynamo theory of terrestrial magnetism 1: Method *Geomag. Aeron.* **11** 585

——1971b Numerical solutions of the equations of the dynamo theory of terrestrial magnetism 2: Results *Geomag. Aeron.* **11** 737

LaBrecque J L, Kent D V and Cande S C 1977 Revised magnetic polarity timescale for Late Cretaceous and Cenozoic time *Geology* **5** 330

Laj C, Nordemann D and Pomeau Y 1979 Correlation function analysis of geomagnetic field reversals *J. Geophys. Res.* **84** 4511

Larson R L 1974 An updated timescale of magnetic reversals for the late Mesozoic *Trans. Am. Geophys. Union* **55** 236

Larson E E, Watson D E and Jennings W 1971 Regional comparison of a Miocene geomagnetic transition in Oregon and Nevada *Earth Planet. Sci. Lett.* **11** 391

Levy E H 1972a Effectiveness of cyclonic convection for producing the geomagnetic field *Astrophys. J.* **171** 621

——1972b Kinematic reversal schemes for the geomagnetic dipole *Astrophys. J.* **171** 635

——1972c On the state of the geomagnetic field and its reversals *Astrophys. J.* **175** 573

Lilley F E M 1970a On kinematic dynamos *Proc. R. Soc.* A **316** 153

——1970b Geomagnetic reversals and the position of the North magnetic pole *Nature* **227** 1336

Lorenz E N 1963 Deterministic non-periodic flow *J. Atmos. Sci.* **20** 130

Lowes F J and Wilkinson I 1963 Geomagnetic dynamo: a laboratory model *Nature* **198** 1158

Lowrie W and Alvarez W 1981 100 million years of geomagnetic polarity history *Geology* **9** 392

Lowrie W and Kent D V 1983 Geomagnetic reversal frequency since the Late Cretaceous *Earth Planet. Sci. Lett.* **62** 305

McElhinny M W 1971 Geomagnetic reversals during the Phanerozoic *Science* **172** 157

McElhinny M W and Merrill R T 1975 Geomagnetic secular variation over the past 5 Myr *Rev. Geophys. Space Phys.* **13** 687

McFadden P L and McElhinny M W 1982 Variations in the geomagnetic dipole 2: statistical analysis of VDMs for the past 5 million years *J. Geomag. Geoelec.* **34** 163

McFadden P L and Merrill R T 1984 Lower mantle convection and geomagnetism *J. Geophys. Res.* submitted

Malkus W V R 1972 Reversing Bullard's dynamo *Trans. Am. Geophys. Union* **53** 617

Mathews J H and Gardner W K 1963 Field reversals of 'Palaeomagnetic' type in coupled disk dynamos *US Naval Res. Lab. Rep.* 5886

Merrill R T and McElhinny M W 1977 Anomalies in the time-averaged palaeomagnetic field and their implications for the lower mantle *Rev. Geophys. Space Phys.* **15** 309

Merrill R T, McElhinny M W and Stevenson D J 1979 Evidence for long term asymmetries in the Earth's magnetic field and possible implications for dynamo theories *Phys. Earth Planet. Int.* **20** 75

Nagata T 1969 Length of geomagnetic polarity intervals *J. Geomag. Geoelec.* **21** 701

Naidu P S 1971 Statistical structure of geomagnetic field reversals *J. Geophys. Res.* **76** 2649

——1974 Are geomagnetic field reversals independent? *J. Geomag. Geoelec.* **26** 101

——1975 Second-order statistical structure of geomagnetic field reversals *J. Geophys. Res.* **80** 803

——1976 Comment on 'Second-order statistical structure of geomagnetic field reversals' by P S Naidu; by T J Ulrych and R W Clayton: Reply *J. Geophys. Res.* **81** 1034

Ness G, Levi S and Couch R 1980 Marine magnetic anomaly timescales for the Cenozoic and Late Cretaceous: a précis, critique and synthesis *Rev. Geophys. Space Phys.* **18** 753

Nozières P 1978 Reversals of the Earth's magnetic field: an attempt at a relaxation model *Phys. Earth Planet. Int.* **17** 55

Parker E N 1969 The occasional reversal of the geomagnetic field *Astrophys. J.* **158** 815

Phillips J D 1977 Time variation and asymmetry in the statistics of geomagnetic reversal sequences *J. Geophys. Res.* **82** 835

Phillips J D, Blakely R J and Cox A 1975 Independence of geomagnetic polarity intervals *Geophys. J.* **43** 747

Phillips J D and Cox A 1976 Spectral analysis of geomagnetic reversal timescales *Geophys. J.* **45** 19

Reid A B 1972 A palaeomagnetic study of 1500 million years in Canada *PhD Thesis* University of Alberta

Rikitake T 1958 Oscillations of a system of disk dynamo *Proc. Camb. Phil. Soc.* **54** 89

Robbins K G 1977 A new approach to sub-critical instability and turbulent transitions in a simple dynamo *Math. Proc. Camb. Phil. Soc.* **82** 309

Roberts P H 1978 Diffusive instabilities in magnetohydrodynamic convection *Les instabilités hydrodynamiques en convection libre, forcés et mixté* (*Lecture Notes in Physics* vol 72) ed J C Legras and J K Platten

Schouten H and Denham C R 1979 Modelling the oceanic magnetic source layer *Deep drilling in the Atlantic ocean: Ocean Crust* ed H Talwani, C G Harrison and D E Hayes *AGU Maurice Ewing Series* **2** 151

Simpson J F 1966 Evolutionary pulsations and geomagnetic polarity *Bull. Geol. Soc. Am.* **77** 197

Sommerville R C J 1967 *Woods Hole Oceanographic Inst. Rep.* No 67–54, vol 2, p 132

Stevenson A F and Wolfson S J 1966 Calculations of the dynamo problem of the Earth's magnetic field *J. Geophys. Res.* **71** 4446

Stewart A D and Irving E 1973 Palaeomagnetism of Precambrian red beds from NW Scotland *Trans. Am. Geophys. Union* **54** 248

Tacier J-D, Switzer P and Cox A 1975 A model relating undetected geomagnetic polarity intervals to the observed rate of reversals *J. Geophys. Res.* **80** 4446

Ulrych T J 1972 Maximum entropy power spectrum of truncated sinusoids *J. Geophys. Res.* **77** 1396

Ulrych T J and Clayton R W 1976 Comment on 'Second-order statistical structure of geomagnetic field reversals' by P S Naidu *J. Geophys. Res.* **81** 1033

Verosub K L 1975a Alternative to the geomagnetic self-reversing dynamo *Nature* **253** 707

——1975b A method for determining a generalised representation of geomagnetic field sources *Geophys. J.* **41** 127

Vogt P R, Einwich A and Johnson G L 1972 A preliminary Jurassic and Cretaceous reversal chronology from marine magnetic anomalies in the western North Atlantic *Trans. Am. Geophys. Union* **53** 363

Watanabe H 1981 Non-steady state of a hydromagnetic $\alpha\omega$-dynamo and its application to the geomagnetic reversals *J. Geomag. Geoelec.* **33** 531

Wells J M 1973 Non-linear spherical harmonic analysis of palaeomagnetic data *Meth. Computat. Phys.* **13** 239

Wilson R L 1970 Permanent aspects of the Earth's non-dipole magnetic field over Upper Tertiary times *Geophys. J.* **19** 417

——1971 Dipole offset—the time averaged palaeomagnetic field over the past 25 million years *Geophys. J.* **22** 491

——1972 Palaeomagnetic differences between normal and reversed field sources, and the problem of far-sided and right-handed pole positions *Geophys. J.* **28** 295

Wilson R L and Ade-Hall J M 1970 Palaeomagnetic indications of a permanent aspect of the non-dipole field *Palaeogeophysics* ed S K Runcorn p 307 (New York: Academic)

Wilson R L and McElhinny M W 1974 Investigations of the large scale palaeomagnetic field over the past 25 Myr: eastward shift of the Icelandic spreading ridge *Geophys. J.* **39** 571

6 Magnetostratigraphy

6.1 Introduction

Magnetostratigraphy organises strata according to their magnetic properties acquired at the time of deposition. Because the polarity of the Earth's magnetic field has reversed repeatedly in the geologic past and because polarity transitions, while lasting only a few thousand years, are synchronous over the entire globe, their record in marine or land-based sediments provides isochrons applicable to worldwide correlation. Khramov (1957) recognised the value of palaeomagnetism in the subdivision and correlation of sedimentary sequences in the Soviet Union. However, it was not until it was used in deep-sea sedimentary sequences many years later that the value of magnetic stratigraphy was fully appreciated. Magnetic reversals are distinct from most other stratigraphic criteria which are characteristically diachronous. There are other synchronous events such as tephra layers which may be used for correlation but which are too restricted geographically for wide-scale use.

Magnetostratigraphy for the last 5 Myr, based on radiometric dating of sequences of terrestrial lava flows, has provided information with which to establish age relationships of sediment and fossil sequences. This chronology in turn has provided a basis for determining rates of geologic change with fair accuracy. It must be stressed, however, that a magnetostratigraphic sequence by itself does not provide unequivocal dates for geological events preserved in sediments, since magnetic reversals are repetitive events and do not in general possess singular properties. Independent criteria are thus required to verify the age of a sequence. However examination of microfossils at only a few stratigraphic levels may enable a biostratigrapher to establish the age of a magnetic polarity sequence.

The development of a timescale for young sequences of volcanic rocks has been described by a number of people (see e.g. McDougall 1977 for a good review). Until the middle 1970s the polarity timescale was confined to rocks younger than about 4.5 Myr. A major difficulty in extending it to older volcanic and sedimentary sequences is the decreasing resolution of the K–Ar method of dating older rocks. The first use of combined magnetic polarity and K–Ar data was by Rutten (1959), who concluded from studies on lavas from volcanoes near Rome that the present normal magnetic interval has lasted for at least 0.47 Myr, and that an earlier interval of normal polarity possibly existed about 2.4 Myr ago. In the early 1960s a large amount of additional data on the same or closely related lavas was published, leading to the development of a polarity timescale for the last 4 Myr. Cox *et al* (1964) collected all the available data and noted that during the last 3.5 Myr changes in polarity had taken place at irregular intervals, and that, within intervals of predominantly one polarity (called epochs), shorter intervals of opposite polarity (called events), about 0.1 Myr in duration, sometimes occurred[†]. During 1966 many papers incorporating new data and giving compilations of previous results were published. These results confirmed the general outline of the polarity timescale and provided evidence for additional short events.

In compiling a local magnetic polarity stratigraphy from a sequence of superposed rocks, of necessity only a finite number of samples can be taken. It is thus quite likely that some polarity changes will have been missed between adjacent samples. It is usually assumed that the space between adjacent points of like polarity has the same polarity i.e. that no short polarity zones have been missed. However the data could equally well be explained if an even number of reversals had taken place in the interval between the two sampling points. On the other hand, if adjacent sampling points have different polarity, it is usually assumed that one magnetic reversal has occurred midway in the space between them. It is clear that the same record would have been obtained if an odd number of reversals had occurred in this interval.

1966 also saw the successful and accurate correlation by Opdyke *et al* and Ninkovich *et al* of the reversal history recorded on land with magnetic polarity measurements in deep-sea sedimentary cores. The first workers to record reversals in deep-sea sediments were Harrison and Funnell (1964) who demonstrated that the last reversal of the Earth's magnetic field occurred within the Quaternary. Later (1966) Harrison obtained an isochronous correlation in sections more than 3000 km apart, and made the first attempt to correlate these reversals with the polarity timescale. In the same year Ninkovich *et al* examined a set of deep-sea cores from the North Pacific and demonstrated quite clearly the presence of the Jaramillo event in deep-sea sediments. This work was also among the first to estimate the duration of the Jaramillo and Olduvai events and to make estimates of the rates of sediment deposition. It was also the first to intercalibrate correlations using both ash layers and palaeomagnetic reversals. In 1967, Hays and Opdyke extended the magnetostratigraphy of Antarctic sediments to about 3.4 Myr using three long piston cores, and obtained strong evidence for the existence of three short

---

† The terminology of polarity intervals is discussed in §6.5. For the moment the terms epoch and event will be retained.

polarity events within the Gilbert reversed epoch. A radiolarian biostrati-graphy, based on the upward sequential disappearance of several radiolarian species, was dated using the polarity scale and evidence put forward for a possible connection between reversals of the Earth's magnetic field and radiolarian extinctions (see also §7.4).

One of the attractions of magnetostratigraphy is that correlation over long distances using fossils is always difficult even when planktonic groups are used. This is because different water masses within the oceans are marked by distinct planktonic assemblages, and even species that do range over wide areas often exhibit stratigraphic ranges in different oceanic regions. The value of magnetostratigraphy was further demonstrated by Hays *et al* (1969) when they dated and correlated a large number of microfossil events in eastern equatorial Pacific deep-sea cores of Quaternary and Pliocene age. They showed that changes in the fossils used for the differentiation of the Miocene–Pliocene boundary were only 4.5–5 Myr old and occurred near the base of the Gilbert epoch. Up until this time, little and often contradictory radiometric evidence had suggested an age of 9 Myr for this boundary. They were also able to date climatic cycles in the Quaternary and found a number of instances of the extinction of microfossils at the time of reversals (see also §7.3).

Johnson and McGee (1983) have developed statistical models which can be used to test the interpretation of a set of magnetic polarity stratigraphy data or to estimate the age information contained in magnetic polarity data. Their models are based on the assumption that the occurrence of magnetic reversals in the geologic past is given by equation (5.21) and that each palaeomagnetic sampling site recovers the true magnetic polarity for its stratigraphic level (in time). They considered three theoretical models for the spacing of a sampling array within a given time increment—uniform, random and exponential spacing. From a knowledge of the number of palaeomagnetic sites, the distribution of the sites, an estimate of the geologic time spanned, and the mean length of the magnetic polarity zones, an estimate of the success of the sampling programme can be made and compared with the observed results. Johnson and McGee estimated the mean timespan for polarity intervals in the late Neogene to be 120 000 years in good agreement with a large body of magnetic polarity stratigraphy data. Once the mean timespan for polarity intervals has been found, an estimate of the time spanned by the stratigraphic section may be made—again they obtained good results for a variety of magnetic polarity sections in the Neogene.

Johnson and McGee further estimated the mean timespan for polarity intervals in the early Palaeogene to be 327 000 years. It must be stressed that their statistical models are based on the spacing of samples in time, not stratigraphic, units. In a typical stratigraphic sequence, unconformities, unsuitable lithologies and differential sedimentation rates will tend to randomise the time spacing. Thus, although a time uniform sampling programme was shown to be the most efficient for establishing a local magnetic polarity stratigraphy, a random sampling model is probably more applicable.

6.2 Epoch boundaries

The age of the Brunhes–Matuyama boundary was determined within narrow limits as 0.69–0.70 Myr by Cox and Dalrymple (1967a), Cox (1969) and Dalrymple (1972). This boundary has become important for the interpolation of ages in deep-sea sedimentary cores, particularly for the detailed interpretation of palaeoclimates from oxygen isotope data (Shackleton and Opdyke 1973). Insufficient reliable polarity and K–Ar data near the boundary between the Matuyama and Gauss epochs led to some uncertainty in fixing its age (2.2–2.5 Myr). Resolution was considerably improved by McDougall and Aziz-ur-Rahman (1972), who found a sequence of normal polarity lavas overlain by flows of reversed polarity on Norfolk Island in the south-west Pacific. They placed the boundary at 2.41 Myr. The age of the boundary between the Gauss and Gilbert epochs was set at 3.32 Myr by Cox (1969) and Dalrymple (1972). There is, however, considerable variation in the estimates of the age of the boundary between the Gilbert epoch and epoch 5, defined on the basis of the polarity history observed in deep-sea cores (Foster and Opdyke 1970). An age of 5.3 ± 0.1 Myr is probably the best estimate at the moment — this is based on direct K–Ar dating and polarity studies on two separate basalt sequences in Iceland by McDougall *et al* (1976a, 1977).

Mankinen and Dalrymple (1979) have published a revised polarity timescale for the past 5 Myr using a significant amount of new data and new values for the atomic abundance and decay activities of ^{40}K (Steiger and Jäger 1977, see table 6.1). Their new values for the major polarity epoch boundaries are 0.73 Myr for the Brunhes–Matuyama, 2.48 Myr for the Matuyama–Gauss, and 3.40 Myr for the Gauss–Gilbert. Their revised polarity timescale is shown in figure 6.1.

Table 6.1 Constants used in the calculation of K–Ar ages.

constant	new values	old values
λ_e (yr^{-1})	0.581×10^{-10}	0.585×10^{-10}
λ_β (yr^{-1})	4.962×10^{-10}	4.72×10^{-10}
$^{40}K/K_{total}$ (mol/mol)	1.167×10^{-4}	1.19×10^{-4}
$^{40}Ar/^{36}Ar$	295.5	295.5

The new values are those recommended by the IUGS Subcommission on Geochronology (Steiger and Jäger 1977).

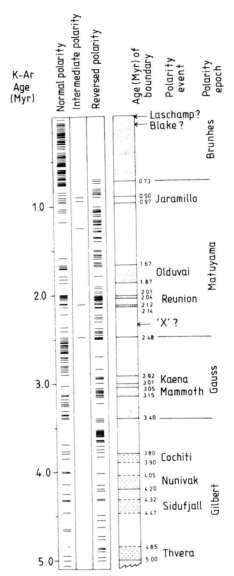

Figure 6.1   Revised Late Cenozoic polarity timescale reconciled to the new K–Ar constants. Each short horizontal line indicates a K–Ar age and magnetic polarity determined for one volcanic cooling unit. The timescale is based primarily on data from table 3a (Mankinen and Dalrymple 1979) but includes information from magnetic profiles and deep-sea sedimentary cores. Stippled pattern indicates periods of normal polarity with coarse stippled pattern indicating events whose limits are poorly defined. Arrows indicate possible brief polarity events (after Mankinen and Dalrymple 1979).

6.3 Age and duration of events

None of the geomagnetic excursions reported from the sedimentary record during the Brunhes epoch are sufficiently worldwide authenticated to be included in the polarity timescale (see chapter 4 for a more detailed discussion). Although there is agreement on the broad aspects of the polarity reversal timescale, considerable uncertainty still exists as to the number, age and duration of normal polarity events in the Matuyama reversed polarity epoch (see e.g. McDougall 1977). The Jaramillo normal polarity event has been estimated by Cox and Dalrymple (1976a) as extending from 0.85–0.97 Myr. Opdyke (1972) placed it between 0.87–0.92 Myr and more recently Mankinen and Dalrymple (1979) between 0.90–0.97 Myr.

The first magnetostratigraphic studies on rocks containing mammal fossils were carried out on volcanic rocks in the Olduvai Gorge area, East Africa by Grommé and Hay (1967, 1971). The volcanic rocks are interbedded with sediments containing rich mammalian assemblages including primitive primates and are important anthropologically. Radiometric dating placed this sequence in the middle of the Matuyama (reversed) epoch and the Olduvai (normal) event at 1.8 Myr. Studies began in the early 1970s in an attempt to correlate the already well known Pliocene–Pleistocene mammalian sequence in the western United States with the late Cenozoic polarity record. The first study was carried out at San Pedro Valley, Arizona by Johnson *et al* (1975) who identified a sequence from the upper Gilbert epoch through to the Brunhes epoch. Four mammal datum planes were palaeomagnetically dated, including the first appearance of the hare *Lepus* at 1.9 Myr marking the appearance of a definitive mammal assemblage. A similar study was carried out in the Anza Borrego State Park by Opdyke *et al* (1977) on a sequence which extended from the Late Gilbert epoch to the Matuyama epoch. The polarity record enabled eight faunal events to be dated including the appearance and extinction of a variety of mammals. The appearance of some of the mammals is almost certainly linked with the Bering Sea land bridge which developed during low stands of sea level and which is thus inferred to have been exposed as a migratory route about 1.9 Myr ago.

There has been considerable controversy as to whether the Olduvai and Gilsa normal events are separate or one and the same. Grommé and Hay (1971) suggested that the two events were not distinct—the lavas from which the Gilsa event was defined having been erupted during the Olduvai event. Data from deep-sea sediment cores that penetrate the Matuyama are equivocal (Watkins 1972, Opdyke 1972) and raise the question of the preservation of short-term events in deep-sea sediments and the causes of spurious polarity reversals (Watkins 1968). In some cores two normal events are seen, although in the majority of cores only a single normal interval can be identified and there has been much argument as to whether this represents the Olduvai or Gilsa event (Watkins 1972). Opdyke (1972) correlated it with the

Olduvai event and, assuming uniform sedimentation rates throughout the Matuyama epoch, estimated it to last from 1.71–1.86 Myr ago. Using all available K–Ar data from the type sequence at Olduvai Gorge, Brock and Hay (1976) showed that nine samples, stratigraphically in about the middle of the normal polarity interval, have concordant K–Ar ages yielding a mean of 1.79 ± 0.03 Myr—in good agreement with the estimates of Opdyke (1972) from deep-sea cores. Watkins *et al* (1975) carried out a detailed palaeomagnetic survey of rocks from the Jokuldalur area, Iceland, covering this period. Their work showed that there is no second normal polarity event in the sections examined. The age of the lavas (based on K–Ar dating) is 1.58 ± 0.08 Myr, which is significantly younger than ages reported for the Olduvai event from other localities. The most recent work of Mankinen and Dalrymple (1979) places the Olduvai event at 1.67–1.87 Myr BP.

The number and duration of the Reunion events are also still uncertain, although available evidence suggests that there are probably two (Cox 1969, Grommé and Hay 1971) and possibly three (Shuey *et al* 1974). K–Ar ages for normally magnetised lavas from different localities give ages between 2.00 and 2.14 Myr (Grommé and Hay 1971). Mankinen and Dalrymple (1979) place two Reunion events at 2.01–2.04 and 2.12–2.16 Myr BP. Only one event is clearly seen in the sequences most recently sampled on Reunion Island (McDougall and Watkins 1973), although it was not possible to correlate these lavas directly with those from earlier studies. The normal interval seen in this latest study has a mean age of 2.07 ± 0.02 Myr with an estimated duration of 0.1–0.05 Myr.

The possible existence of a normal polarity event in the Matuyama reversed epoch older than the Reunion events, identified as anomaly X in magnetic anomaly profiles by Heirtzler *et al* (1968) and Emilia and Heinrichs (1969), has not been confirmed by land-based studies. It has been seen in other magnetic anomaly profiles (Blakely and Cox 1972, Rea and Blakely 1975) and in a sedimentary sequence (Shuey *et al* 1974) but has not been detected in a volcanic sequence with the possible exception of a 2.37 Myr normal polarity lava from Argentina (Valencio *et al* 1970).

Two reversed polarity events have been found in the Gauss normal polarity epoch, the Mammoth event, defined and dated at 3.06 ± 0.10 Myr by Cox *et al* (1964), and the Kaena event first defined and dated by McDougall and Chamalaun (1966). Later McDougall and Aziz-ur-Rahman (1972), using more recent data including results from deep-sea cores, concluded that the Kaena event lasted from 2.92–2.84 Myr. They also estimated that the Mammoth reversed interval occurred between 2.99 and 3.09 Myr ago. Mankinen and Dalrymple (1979) place the Kaena event at between 2.92 and 3.01 Myr and the Mammoth between 3.05 and 3.15 Myr.

Liddicoat *et al* (1980) have examined the palaeomagnetic polarity in a 930 m core from Searles Valley, California (see figure 6.2). For 320 m below a depth of 185 m the inclination is predominantly negative and Liddicoat *et al* assume it

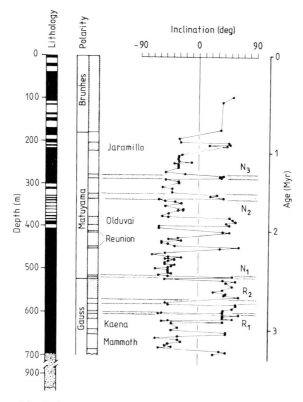

Figure 6.2 Palaeomagnetic polarity in Searles Valley, California. Shaded region highlights the unnamed zones of inclination, three in the Matuyama epoch and two in the Gauss epoch. In the lithological column: unshaded area, salines; black area, mud; spotted area, arkose. In the polarity column: hatched area, normal polarity; unshaded area, reversed polarity. Age is extrapolated from polarity boundaries in the timescale of Mankinen and Dalrymple (1979), and by assuming a uniform sedimentation rate. Note that the timescale is not linear (after Liddicoat *et al* 1980).

to represent the Matuyama epoch. Below that the inclination is predominantly positive, which they assume to be the upper part of the Gauss epoch. Assuming that the Brunhes–Matuyama and Matuyama–Gauss boundaries have been correctly recognised, it is possible to plot age against depth to identify the zones of positive inclination within the Matuyama. In this way Liddicoat *et al* identified the Jaramillo, Olduvai and two Reunion events. Three other short segments of positive inclination, not included in the Mankinen and Dalrymple timescale (1979) are also present in the record (N_1, N_2, N_3 in figure 6.2). Liddicoat *et al* speculate that N_3 might be the same as that seen by Mankinen *et al* (1978) at Cobb Mountain, California, by Watkins and Abdel-Monem (1971) in a basalt from Madeira, by Briden *et al* (1979) in a basalt from the Lesser Antilles (Guadaloupe and St Vincent) or by Maenaka (1979) in a normally magnetised ash layer in the Osaka Group, Japan. Mankinen and Grommé (1982) have since found two further cases of normal polarity in some basalt flows which erupted in the Coso Range, California about 1.08 Myr ago, about the same time as the Cobb Mountain normal event which they had reported earlier (1978). Reviewing all the available data they came to the conclusion that a worldwide geomagnetic event (a complete reversal of polarity, not just an excursion of the VGP to low latitudes) occurred just prior to the Jaramillo normal polarity event. It seems that the Cobb Mountain normal polarity event lasted only for about 10 000 years. Precise estimates of its duration are difficult to obtain since such short lengths of time are beyond the resolution of radiometric dating. The N_2 event may correlate with the controversial Gilsa event. Finally the N_1 event might correspond to the X event (Heirtzler *et al* 1968) seen in some magnetic anomaly data or to that observed in Icelandic lavas very close to the Matuyama–Gauss boundary (Kristjansson *et al* 1978).

In the Gauss epoch, Liddicoat *et al* identified the Kaena and Mammoth events and in addition found two other short segments of negative inclination (R_1, R_2 in figure 6.2). Some reversely magnetised K–Ar dated basalt flows in the age range of R_1, R_2 have been reported—two from Oaha in Hawaii (Doell and Dalrymple 1973), one in west Iceland (McDougall *et al* 1977) and one in St Vincent (Briden *et al* 1979). The advantage of using lacustrine deposits such as those in Searles Valley to look for short events is that the sedimentation rate is at least fifty times greater than that in most deep-sea sediments and is much less discontinuous than the eruption of lava flows.

Cox and Dalrymple (1976b) identified two normal polarity intervals within the Gilbert reversed epoch—these were named the Cochiti event, dated at 3.7 Myr, and the Nunivak event, dated at 4.1 Myr. These events are regarded as the same as events 'a' and 'b' seen in deep-sea cores by Hays and Opdyke (1967). Two older events have also been found in marine sediments and called C_1 and C_2 by Foster and Opdyke (1970) with mean ages of about 4.37 and 4.56 Myr (Opdyke 1972). Both these older intervals of normal polarity in the Gilbert epoch have subsequently been identified in a continuous sequence of lava flows in western Iceland by McDougall *et al* (1977) and named the Sidufjall and Thvera events. The ages of the boundaries of the Sidufjall event were estimated to be 4.33 and 4.45 Myr and those of the Thvera event 4.60 and 4.80 Myr. Mankinen and Dalrymple (1979) place these two events at 4.32–4.47 Myr and 4.85–5.00 Myr. A slightly different Quaternary polarity time-scale has been given by Berggren *et al* (1980).

6.4 Extension of the polarity timescale

The hope of extending the polarity timescale to times earlier than 5 Myr using K–Ar dating and polarity measurements on stratigraphically unrelated samples is not very promising. This is because of the lack of resolution in K–Ar dating—the best K–Ar dates are subject to errors of about 5% so that it is difficult to define a magnetic transition from only one determination. Cox and Dalrymple (1967a) have shown that a greater precision may be obtained if a number of measurements are made on lavas near a magnetic boundary. The same difficulties exist in extending the polarity timescale beyond 5 Myr to try to detect additional reversals. However, Watkins and Walker (1977), Mc-Dougall *et al* (1976b, 1977) and Harrison *et al* (1979) have had some success in extending the timescale back to 13.6 Myr. They used volcanic sequences in Iceland where there are thick sequences of lavas representing significant amounts of time and where stratigraphic relationships between successive lavas are clear.

Figure 6.3 shows the results of magnetic polarity and K–Ar dating of a 3400 m thick succession of basaltic lavas in the Borgarfjordur region in western Iceland. A considerable number of polarity changes were recorded

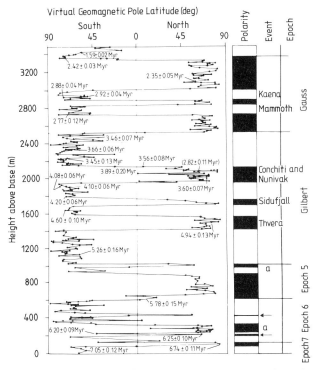

Figure 6.3   Plot of latitude of vGP against stratigraphic height above the base of a 3500 m thick succession of basaltic lavas from western Iceland. Interpretation of the data in terms of polarity given below—black indicates normal polarity, white, reversed polarity (after McDougall *et al* 1977).

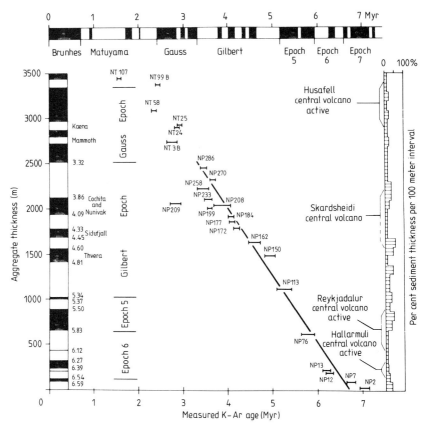

Figure 6.4   K–Ar ages, with precision limits plotted against aggregate stratigraphic thickness above the base of a 3500 m thick sequence of basaltic lavas from western Iceland. Polarity log for the sequence shown adjacent to the thickness axis; black indicates normal polarity, white, reversed polarity. Polarity timescale after Talwani *et al* (1971) given at top of diagram. Regression line through data is least-squares fit with Gilbert–Gauss epoch boundary fixed at 3.32 Myr at 2520 m. Ages indicated for polarity interval boundaries on polarity log derived from regression. The proportion of sediment present for each 100 m of section shown on right, and period of activity of central volcanoes in the region is also indicated (after McDougall *et al* 1977).

from the early Matuyama into epoch 7. Figure 6.4 shows K–Ar ages plotted against stratigraphic height above the base of the sequence—a least-squares fit regression line enables ages to be obtained for the polarity interval boundaries. Figure 6.5 shows excellent agreement between these results and extrapolations based on marine magnetic anomaly data—particularly with those of Talwani *et al* (1971) and Klitgord *et al* (1975). Figure 6.6 shows the results obtained by

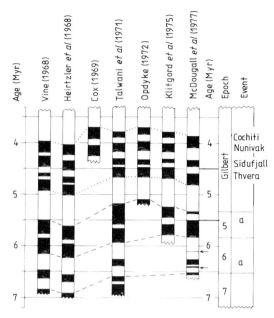

Figure 6.5   Comparison of polarity timescale derived by McDougall *et al* (1977) with scales proposed by other workers. Black indicates normal polarity, white, reversed polarity (after McDougall *et al* 1977).

McDougall *et al* (1976b) from part of a 2330 m thick lava sequence in the Neskaupstadur region of eastern Iceland where they were able to extend the polarity timescale back to over 12 Myr. Again there is good agreement between the observed polarity in the lava succession in Iceland and marine magnetic anomaly data. Unless similar thick sequences of lavas of the right age can be found it is unlikely that further extension of the polarity timescale by direct K–Ar dating and polarity measurements can be made. The only recourse is to use marine magnetic anomaly data as Heirtzler *et al* first did in 1968 when they produced a timescale going back 76 Myr. Such a timescale involves an extrapolation of the order of 20 times the length of the base line and there is a real possibility of progressively larger errors with increasing age. However, the work in Iceland does give some confidence in using marine data to extend the polarity timescale to earlier times.

Because of drilling disturbance some of the cores obtained in the Deep Sea Drilling Project (DSDP) could not be used directly for magnetostratigraphic purposes. Calibration of the polarity history with biostratigraphy of these deep-drilled sedimentary sequences has had to be carried out indirectly by radiometric dating of planktonic microfossil datums of marine sections exposed on land. Using two long piston cores from the equatorial Pacific with low sedimentation rates, Foster and Opdyke (1970) were able to extend the

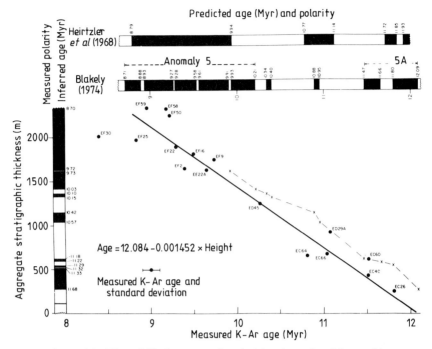

Figure 6.6 Plot of K–Ar age against height above local base of lava sequence in Neskaupstadur region of eastern Iceland. Regression line through data points is indicated and ages of boundaries between polarity intervals given on polarity log for the sequence at left are derived from the regression. Polarity logs based upon analysis of marine magnetic anomalies by Heirtzler *et al* (1968) and Blakely (1974) are given at the top of the diagram. Black indicates normal polarity, white, reversed polarity. The broken line is obtained by fitting Blakely's dates (indicated by crosses) to the eastern Iceland polarity interval boundaries (after McDougall *et al* 1976b).

magnetic stratigraphy in deep-sea sediments back to about 12 Myr. Their results are in substantial agreement with the pattern of polarity epochs predicted by sea-floor spreading. They did not calibrate their magnetic record with any biostratigraphic data because of a lack of evidence of stratigraphic hiatuses and similarities between the magnetostratigraphic records in the two cores. Reviews of palaeomagnetism of deep-sea sediments have been given by Opdyke (1972), Harrison (1974) and Butler and Opdyke (1979). McKee and Elston (1980) studied the reversal pattern in a 7.9–11.5 Myr volcanic sequence from the Hackberry Mountain area in central Arizona and compared it with the ocean floor polarity record. Although there is general agreement, they found no detailed correlations, the Hackberry Mountain pattern comparing less well with the marine sedimentary core chronology of Opdyke (1972) and

Opdyke *et al* (1974) than with the ocean floor anomaly record of LaBrecque *et al* (1977). This implies that there are unrecognised hiatuses in the marine sediment record. The lack of correlation between the different polarity chronologies reflects the difficulty of accurate dating, the loss or absence of the geologic record in the terrestrial volcanic sequence, assumptions about the uniformity of the rates of sea-floor spreading and the accumulation of volcanic and sedimentary rocks, and the lack of fine-scale resolution of the magnetic record of ocean floor anomalies so that brief reversals in the marine record may be missed. Radiometric dates on rocks older than about 6.0 Myr have an uncertainty greater than the length of many magnetic reversals. Besides fluctuations in continuous deposition, unconformities exist that represent variable periods of non-deposition, erosion or both. It is impossible to estimate the amount of time represented by such unconformities. A change in magnetic polarity in two stratigraphic adjacent units gives no information about the age difference between the units. A series of reversals also tells little about the total amount of time represented by, or missing from, the record.

The next development was the extension of the polarity timescale to marine sedimentary sections uplifted on land. One of the first investigations was carried out by Kennett *et al* (1971) on a Pliocene to early Quaternary section at Mangaopari Stream in the North Island of New Zealand. The principal significance of this work was its demonstration that magnetostratigraphy could be applied to land-based marine sections and hence help in establishing the chronology of late Cenozoic climatic history. Kennett (1980) has cautioned against the uncritical acceptance of the magnetostratigraphy from land-based marine sections since several factors may complicate the record, particularly normal palaeomagnetic overprinting resulting from post-depositional chemical changes in the sediment. The situation greatly improved during the 1970s with the development of more rapid and sensitive magnetometers, digital spinners and, later, cryogenic instruments making possible the measurement of large numbers of weakly magnetised samples. In 1977 a series of papers was published on the geomagnetic record preserved in marine sediments at Gubbio, Italy (Arthur and Fischer 1977, Premoli Silva 1977, Lowrie and Alvarez 1977a, Roggenthen and Napoleone 1977, Alvarez *et al* 1977). A long, continuous sequence of magnetic polarity zones was observed in sediments of Middle Cretaceous to Palaeocene age, which were intercalibrated with planktonic foraminiferal zones. The Gubbio section is particularly valuable since it contains an unbroken sequence across the Cretaceous–Tertiary boundary (Luterbacher and Premoli Silva 1964). In most of the documented Late Cretaceous–Early Tertiary sections of the world, the boundary between the periods is represented by a hiatus. The magnetostratigraphy at Gubbio also closely matches the polarity sequence inferred from marine magnetic anomaly profiles. This is confirmed by a series of further biostratigraphic and magnetic stratigraphic investigations of late Mesozoic and Cenozoic pelagic carbonate rocks (see e.g. Alvarez and Lowrie 1978, Channell *et al* 1979, Lowrie *et al* 1980a, b, 1982, Vandenberg and Wonders 1980, Napoleone *et al* 1983). The combined data from Lowrie and Alvarez (1977b) and Napoleone *et al* (1983) from the Bottaccione Gorge near Gubbio are shown in figure 6.7.

Work has also been carried out on the magnetostratigraphy of non-marine sediments. Helsley (1969) sampled the Early Triassic, largely unfossiliferous, Moenkopi Formation of the Colorado Plateau and found at least eleven reversals. He attempted to correlate his results with the reversal sequence reported by Picard (1964) from the Chugwater Formation, Wyoming, and also with the anomaly patterns observed at sea. This work was later extended to younger parts of the Moenkopi Formation by Helsley and Steiner (1974) who confirmed the mixed polarity character of the early Triassic. Similar studies were carried out independently by Burek (1970) on the Triassic Upper Buntsandstein of south-west Germany. It would be extremely valuable to be able to correlate the North American and European sequences— unfortunately this has not yet proved possible using magnetostratigraphy because, in the absence of fossils, there is no way of separating magnetic polarity changes from one another. Both Burek (1970) and Helsley and Steiner (1974) emphasised the importance of the differences between the Early Triassic mixed polarity record and the long periods of constant polarities in both the underlying Permian and overlying late Triassic to Cretaceous sediments. A fundamental change in the polarity history occurred at the end of the late Cretaceous when the frequency of reversals increased. This was followed by a further increase in the frequency of reversals about 45 Myr ago (see §3.5). Helsley and Steiner suggested that major geologic eras, defined on the basis of major changes in the fossil record, may end with a low frequency of reversals and begin with a high reversal frequency. They speculated that major changes in life forms occurred at times of rapid polarity changes (see also §7.4).

The magnetic polarity timescale of Heirtzler *et al* (1968) was derived on the assumption that the magnetic anomaly profiles seen in the oceans were manifestations of earlier reversals of the Earth's magnetic field and on a constant rate of sea-floor spreading in the South Atlantic back to about 80 Myr. The extrapolation from a calibration date of 3.55 Myr was supported at the time it was made by data from a single Late Cretaceous core from the South Atlantic near anomaly 31. There have been many new versions of the polarity timescale since then, yet for the most part they are fairly small revisions, additions or recalibrations (for a good review of the subject up to 1979, see Ness *et al* 1980). Biostratigraphic results from the DSDP have confirmed the general accuracy of such timescales and have also been used to calibrate portions of them. The disadvantage of the original Heirtzler polarity timescale is that it was generated from a single sinuous profile obtained from the relatively slow-spreading South Atlantic Ridge, which affects both its resolving power and accuracy. Revised timescales have been proposed by LaBrecque *et al* (1977), Ness *et al* (1980), Lowrie and Alvarez (1981) and Butler and Coney (1981). LaBrecque *et al* raised the question of small scale anomalies

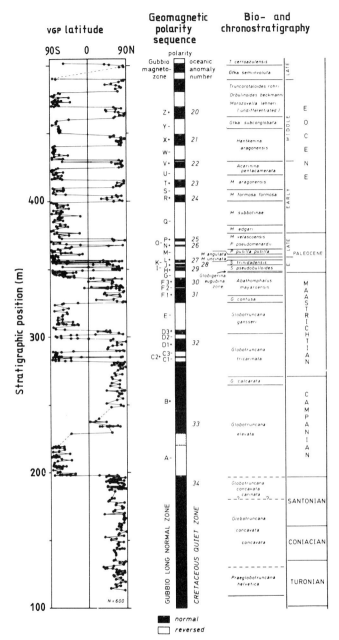

Figure 6.7 Complete biostratigraphy and magnetic stratigraphy of the Scaglia limestones in the Bottaccione Gorge near Gubbio, Italy, showing the identification and dating of marine magnetic anomalies associated with the derived geomagnetic polarity sequence. The data are composited from Lowrie and Alvarez (1977b) and Napoleone *et al* (1983) (after Lowrie and Heller 1982).

('tinywiggles') that have been reported by a number of workers as full-scale reversals. These features have wavelengths of the order of 20 km and amplitudes of 50–80 γ. They suggested that such features may be intensity fluctuations of the Earth's dipole field rather than field reversals (see also §5.5).

The Cenozoic magnetic polarity timescale of LaBrecque *et al* (1977) primarily involved two calibration points—an age of 3.32 Myr for the Gilbert–Gauss boundary and an age of 64.9 Myr for the older boundary of magnetic anomaly 29. The latter calibration point resulted from the placement of the Cretaceous–Tertiary boundary just preceding anomaly 29. From analyses of biostratigraphic ages of DSDP sediments, Berggren *et al* (1978) concluded that magnetic anomaly 24 was basal Eocene rather than late Palaeocene as on the timescale of LaBrecque *et al*. Ness *et al* (1980) proposed a revised polarity timescale using four calibration points: an age of 3.40 Myr for the Gilbert–Gauss boundary, 10.30 Myr for the older boundary of anomaly 5, 54.90 Myr for the older boundary of anomaly 24 consistent with the placement of anomaly 24 as basal Eocene, and an age of 66.70 Myr for the Cretaceous–Tertiary boundary just preceding the older boundary of anomaly 29. For geologic time boundaries in the Palaeocene they used the geologic timescale of Hardenbol and Berggren (1978), with absolute ages recalculated using revised decay constants. In referring to their polarity timescale, Ness *et al* described it as 'at best a critical reshuffling of some very old cards from some very different decks'.

Lowrie and Alvarez (1981) used eleven calibration points in their magnetic polarity timescale. This is based on magnetostratigraphic studies in the Gubbio pelagic limestones, which revealed a reversal sequence of long and short polarity zones which almost exactly matches the marine magnetic anomaly sequence of LaBrecque *et al* (1977) and which has been palaeontologically dated on the basis of abundant foraminifera. These records can be precisely linked because, for each stratigraphic level, the magnetic and palaeontological information both come from the same small sample. Lowrie and Alvarez (1981) were able to fix the positions in the magnetic polarity sequence where various appearances and extinctions of planktonic foraminifera and coccolith taxa occur. The most significant relocation concerns the Palaeocene Eocene boundary which they located within the negative polarity zone just younger than anomaly 25 and the Tertiary–Cretaceous boundary in the negative zone between anomalies 29 and 30. The Palaeocene–Eocene boundary varies between anomalies 24 and 25 in agreement with its location in the Umbrian marine limestones (Lowrie and Alvarez 1981). These data indicate that the older boundary of anomaly 24 is 52.5 Myr. This age is younger than the late Palaeocene age assigned by LaBrecque *et al* (1977) and also younger than the basal Eocene age assigned by Ness *et al* (1980). Butler and Coney (1981) have given a revised magnetic polarity timescale for the Palaeocene and early Eocene using these new data. A comparison of the magnetic polarity timescales of LaBrecque *et al*, Ness *et al* and Butler and

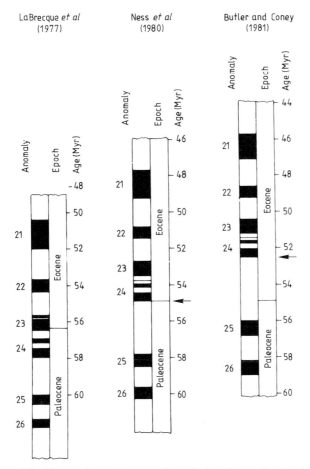

Figure 6.8 Comparison of magnetic polarity timescales in the Palaeocene and early Eocene. Ages of LaBrecque *et al* (1977) timescale have
been corrected as tabulated by Mankinen and Dalrymple (1979). Arrows
to the right of timescales of Ness *et al* (1980) and Butler and Coney (1981)
indicate the calibration point at the older boundary on anomaly 24 used in
constructing these timescales (after Butler and Coney 1981).

Coney is shown in figure 6.8. The ages of magnetic polarity intervals between
anomalies 21 and 24 are about 5 Myr younger in Butler and Coney's timescale
compared to that of LaBrecque *et al*. Schlich (1981) has proposed a further
revision of Heirtzler *et al*'s magnetic polarity timescale for the Eocene and Late
Cretaceous using the results of deep-sea drilling by the *Glomar Challenger*. His
largest difference is 7.5 Myr and is located near anomalies 25 and 26 in the Late
Palaeocene.

A revised Palaeogene geochronology has been put forward by Berggren *et al*
(1983). An assessment of first order correlations of calcareous plankton

biostratigraphic datum events to magnetic polarity stratigraphy gives the following estimated 'magnetobiochronology': Cretaceous–Tertiary boundary (just below anomaly 29): 66.4 Myr; Palaeocene–Eocene (between anomalies 24 and 25): 57.8 Myr; Eocene–Oligocene (between anomalies 13 and 15): 36.6 Myr; Oligocene–Miocene (mid-anomaly 6C): 23.7 Myr.

Larson and Helsley (1975) have reviewed the marine magnetic anomaly and palaeomagnetic record of the Mesozoic. One of the most interesting features is the lack of anomalies in what has been called the Jurassic quiet zone. Lack of anomalies has been interpreted as implying that a major portion of the Middle Jurassic was of constant normal polarity. However, palaeomagnetic studies by Steiner (1978) and Steiner *et al* (1977) of the Curtis, Summerville, Sundance and lower Morrison Formations suggest that Callovian time was dominantly of reversed polarity.

Although much more data are required to explain the Jurassic quiet zone adequately, the previously favoured explanation involving a long period of constant polarity is open to question. Magnetic polarity stratigraphy studies of pre-Jurassic strata are hampered by the lack of an independent record of the reversal sequence. The sequence can only be determined by polarity stratigraphy of sedimentary sequences, with occasional help from palaeomagnetic studies of igneous and metamorphic rocks.

6.5 Magnetostratigraphic polarity units

The purpose of magnetostratigraphic classification is to organise rock strata systematically into identifiable units based on stratigraphic variations in their magnetic characteristics. In considering the chronostratigraphic value of polarity horizons and units, it must be stressed that, although they may be very useful guides to isochronous position, they have relatively little individuality (one reversal being very much the same as another) and therefore can usually only be unequivocally identified by the use of supporting evidence, such as palaeontologic or radiometric data. A magnetostratigraphic polarity unit is an objective unit ideally based on a directly determinable property of the rocks — their magnetic polarity. The presence of the unit can be strictly assured only where this property can be identified. In these respects, it is more similar to a lithostratigraphic or a biostratigraphic unit than to a chronostratigraphic unit. However, lithostratigraphic and biostratigraphic units are usually geographically quite restricted, whereas a polarity unit is potentially worldwide and in this respect more similar to a chronostratigraphic unit. Again it must be emphasised that they are not chronostratigraphic units since they are defined primarily, not by time, but by a specific physical character — the polarity of remanent magnetism.

The best sequential record of reversals of the Earth's magnetic field for the past 150 Myr is preserved in the pattern of sea-floor spreading anomalies and has been dated by extrapolation and interpolation from radiometric and palaeontologic evidence. However, because of the nature of these numbered linear magnetic intensity anomalies from the ocean floor, it is not possible to designate any satisfactory type intervals or type boundaries. Instead, the standards of reference for the marine magnetic anomalies must remain profiles such as those described by Heirtzler *et al* (1968), the boundaries of the units being determined by model fitting.

Ideally, the standard for the definition and recognition of a magnetostratigraphic polarity unit should be a clearly designated stratotype in a continuous sequence of rock strata—a specific section showing the polarity pattern of the unit throughout and clearly defining its upper and lower limits by means of boundary stratotypes. Alvarez *et al* (1977) have attempted to establish a stratotype section of this sort in the Gubbio section of Italy where Upper Cretaceous polarity units have been identified geologically and geographically and related to lithostratigraphic and biostratigraphic data from this section.

Rock magnetic polarity units have been established in two different ways: (1) through a combination of radiometric or biostratigraphic age data with magnetic polarity determinations on outcropping or cored volcanic and sedimentary rocks, and (2) through the use of magnetometer profiles from ocean surveys to identify and correlate linear magnetic anomalies that are interpreted as reversals of the Earth's magnetic field recorded in the lavas of the sea floor during sea-floor spreading, usually in the complete absence of any direct radiometric data. Any system should be flexible enough to allow for polarity intervals discovered at a later date to be conveniently and unambiguously inserted into the existing system. The first four periods of dominantly one polarity were called epochs, and named after past workers in geomagnetism (shorter intervals of one polarity that occurred during the above epochs were named after the locations where they were first observed)[†]. Older palaeomagnetic epochs have been numbered in a similar manner to the numerical system developed by workers at Lamont during the 1960s for sea-floor magnetic anomalies. Unfortunately, the numbers that identify the different polarity events do not coincide on the two systems. By the early 1970s, inconsistencies and nomenclature conflicts in magnetic stratigraphy were becoming more serious, and in 1972 a subcommission on a Magnetic Polarity Time Scale (SMPTS) was established as part of the International Commission on Stratigraphy, which in turn is a part of the International Union of Geological Sciences. After a series of meetings, this commission produced a set of recommendations in an attempt to establish an unambiguous nomenclature in magnetostratigraphy, which is as consistent as possible with conventional stratigraphic terminology. Recommendations by the subcommission, which were rather general, were published in 1973 and guidelines were given by Watkins (1976).

[†] These names should be preserved, but should be considered to be primarily polarity zones and not formal chronostratigraphic and geochronological units.

Table 6.2

Magnetostratigraphic polarity units	Geochronological equivalent	Chronostratigraphic equivalent
Polarity superzone	Chron (or superchron)	Chronozone (or super chronozone)
Polarity zone	Chron	Chronozone
Polarity subzone	Chron (or subchron)	Chronozone (or sub-chronozone)

The report however, failed to satisfy fully the International Subcommission on Stratigraphic Classification (ISSC) who in turn produced a document on magnetostratigraphic polarity units (1978), which was then reviewed by members of the SMPTS. The resulting report was published as a supplementary chapter of the International Stratigraphic Guide and is the prepared statement of both the ISSC and the SMPTS. The most significant change in terminology adopted is the replacement of the word 'epoch', as in Brunhes epoch, with the word 'chron'. The expression Matuyama chron, or Matuyama polarity chron, is thus preferable to Matuyama epoch. Furthermore, the word 'event', as in Jaramillo event, is replaced by the word 'subchron', as in Jaramillo subchron (of the Matuyama chron). These are chronological (geochronological) units. The chronostratigraphic equivalents for these terms as adopted are 'chronozone' for 'chron' and 'subchronozone' for 'subchron'. The recommended terminology for magnetostratigraphic polarity units and their geochronological and chronostratigraphic equivalents is given in table 6.2.

References

Alvarez W, Arthur M A, Fischer A G, Lowrie W, Napoleone G, Premoli Silva I and Roggenthen W M 1977 Upper Cretaceous–Palaeocene magnetic stratigraphy at Gubbio, Italy. V. Type section for the late Cretaceous–Palaeocene geomagnetic reversal timescale *Geol. Soc. Am. Bull.* **88** 383

Alvarez W and Lowrie W 1978 Upper Cretaceous palaeomagnetic stratigraphy at Moria (Umbrian Apennines, Italy): verification of the Gubbio section *Geophys. J.* **55** 1

Arthur M A and Fischer A G 1977 Upper Cretaceous–Palaeocene magnetic stratigraphy at Gubbio, Italy. I. Lithostratigraphy and sedimentology *Geol. Soc. Am. Bull.* **88** 367

Berggren W A, Burckle L H, Cita M B, Cooke H B S, Funnell B M, Gartner S, Hays J D, Kennett J P, Opdyke N D, Pastouret L, Shackleton N J and Takayanagi Y 1980 Towards a Quaternary timescale *Quat. Res.* **13** 277

Berggren W A, Kent D V and Flynn J J 1983 Paleogene geochronology and chronostratigraphy *Geochronology and the Geological Record* ed N J Snelling

(London: Geol. Soc. London, Special Paper)

Berggren W A, McKenna M C, Hardenbol J and Obradovich J D 1978 Revised Palaeocene polarity timescale *J. Geol.* **86** 67

Blakely R J 1974 Geomagnetic reversals and crustal spreading rates during the Miocene *J. Geophys. Res.* **79** 2979

Blakely R J and Cox A 1972 Identification of short polarity events by transforming marine magnetic profiles to the pole *J. Geophys. Res.* **77** 4339

Briden J C, Rex D C, Faller A M and Tomblin J F 1979 K–Ar geochronology and palaeomagnetism of volcanic rocks in the Lesser Antilles island arc *Phil. Trans. R. Soc.* A **291** 485

Brock A and Hay R L 1976 The Olduvai event at Olduvai Gorge *Earth Planet. Sci. Lett.* **29** 127

Burek P J 1970 Magnetic reversals: their application to stratigraphic problems *Am. Assoc. Pet. Geol. Bull.* **54** 1120

Butler R F and Coney P J 1981 A revised magnetic polarity timescale for the Palaeocene and early Eocene and implications for Pacific plate motion *Geophys. Res. Lett.* **8** 301

Butler R F and Opdyke N D 1979 Magnetic polarity stratigraphy *Rev. Geophys. Space Phys.* **17** 235

Channell J E T, Lowrie W and Medizza F 1979 Middle and Early Cretaceous magnetic stratigraphy from the Cismon section, northern Italy *Earth Planet. Sci. Lett.* **42** 153

Cox A 1969 Geomagnetic reversals *Science* **163** 237

Cox A and Dalrymple G B 1967a Statistical analysis of geomagnetic reversal data and the precision of potassium–argon dating *J. Geophys. Res.* **72** 2603

——1967b Geomagnetic polarity epochs—Nunivak Island, Alaska *Earth Planet. Sci. Lett.* **3** 173

Cox A, Doell R R and Dalrymple G B 1964 Reversals of the Earth's magnetic field *Science* **144** 1537

Dalrymple G B 1972 Potassium–argon dating of geomagnetic reversals and North American glaciations *Calibration of Hominoid Evolution* ed W W Bishop and J A Miller (Edinburgh: Scottish Acad. Press)

Doell R R and Dalrymple G B 1973 Potassium–argon ages and palaeomagnetism of the Waianae and Koolau volcanic series, Oahu, Hawaii *Bull. Geol. Soc. Am.* **84** 1217

Emilia D A and Heinrichs D F 1969 Ocean floor spreading: Olduvai and Gilsa events in the Matuyama epoch *Science* **166** 1267

Foster J H and Opdyke N D 1970 Upper Miocene to Recent magnetic stratigraphy in deep-sea sediments *J. Geophys. Res.* **75** 4465

Grommé C S and Hay R L 1967 Geomagnetic polarity epochs: new data for Olduvai Gorge, Tanganyika *Earth Planet. Sci. Lett.* **2** 111

——1971 Geomagnetic polarity epochs: age and duration of the Olduvai normal polarity event *Earth Planet. Sci. Lett.* **10** 179

Hardenbol J and Berggren W A 1978 A new Palaeocene numerical timescale *Contributions to the Geologic Time Scale, Studies of Geology* **6** Am. Assoc. Pet. Geol. **213**

Harrison C G A 1966 The palaeomagnetism of deep-sea sediments *J. Geophys. Res.* **71** 3033

——1974 The palaeomagnetism record from deep-sea sediment cores *Earth Sci. Rev.* **10** 1

Harrison C G A and Funnell B M 1964 Relationship of palaeomagnetic reversals and micropalaeontology in two late Cenozoic cores from the Pacific Ocean *Nature* **204** 566

Harrison C G A, McDougall I and Watkins N D 1979 A geomagnetic field reversal timescale back to 13.0 million years before present *Earth Planet. Sci. Lett.* **42** 143

Hays J D and Opdyke N D 1967 Antarctic radiolaria, magnetic reversals and climatic change *Science* **158** 1001

Hays J D, Saito T, Opdyke N D and Burckle L H 1969 Pliocene–Pleistocene sediments of the equatorial Pacific; their palaeomagnetic, biostratigraphic, and climatic record *Geol. Soc. Am. Bull.* **80** 1481

Heirtzler J R, Dickson G O, Herron E M, Pitman W C and Le Pichon X 1968 Marine magnetic anomalies, geomagnetic field reversals, and motions of the ocean floor and continents *J. Geophys. Res.* **73** 2119

Helsley C E 1969 Magnetic reversal stratigraphy of the Lower Triassic Moenkopi formation of Western Colorado *Geol. Soc. Am. Bull.* **80** 2431

Helsley C E and Steiner M B 1974 Palaeomagnetism of the lower Triassic Moenkopi formation *Bull. Geol. Soc. Am.* **85** 457

Johnson N M and McGee V E 1983 Magnetic polarity stratigraphy: stochastic properties of data, sampling problems, and the evaluation of interpretations *J. Geophys. Res.* **88** 1213

Johnson N M, Opdyke N D and Lindsay E H 1975 Magnetic polarity stratigraphy of Pliocene–Pleistocene terrestrial deposits and vertebrate faunas, San Pedro Valley, Arizona *Bull. Geol. Soc. Am.* **86** 5

Kennett J P (ed) 1980 *Magnetic stratigraphy of sediments* (Dowden: Hutchingson & Ross).

Kennett J P, Watkins N D and Vella P 1971 Palaeomagnetic chronology of Pliocene–early Pleistocene climates and the Plio–Pleistocene boundary in New Zealand *Science* **171** 276

Khramov A N 1957 Palaeomagnetism: the basis of a new method of correlation and subdivision of sedimentary strata *Dokl. Akad. Nauk. Earth Sci. Sec. Proc.* **112** 129

Klitgord K D, Huestis S P, Mudie J D and Parker R L 1975 An analysis of near bottom magnetic anomalies: sea floor spreading and the magnetised layer *Geophys. J.* **43** 387

Kristjansson L, Fridleifsson I B and Watkins N D 1978 Palaeomagnetism of the Esja area, SW Iceland *Trans. Am. Geophys. Union* **59** 270

LaBrecque J L, Kent D V and Cande S C 1977 Revised magnetic polarity timescale for Late Cretaceous and Cenozoic time *Geology* **5** 330

Larson R L and Helsley C E 1975 Mesozoic reversal sequence *Rev. Geophys. Space Phys.* **13** 174

Liddicoat J C, Opdyke N D and Smith G I 1980 Palaeomagnetic polarity in a 930 m core from Searles Valley, California *Nature* **286** 22

Lowrie W and Alvarez W 1977a Upper Cretaceous–Palaeocene magnetic stratigraphy at Gubbio, Italy. III. Upper Cretaceous magnetic stratigraphy *Bull. Geol. Soc. Am.* **88** 374

——1977b Late Cretaceous geomagnetic polarity sequence: detailed rock and palaeomagnetic studies of the Scaglia Rossa limestone at Gubbio, Italy *Geophys. J.* **51** 561

——1981 One hundred million years of geomagnetic polarity history *Geology* **9** 392

Lowrie W, Alvarez W, Premoli Silva I and Monechi S 1980a Lower Cretaceous magnetic stratigraphy in Umbian pelagic carbonate rocks *Geophys. J.* **60** 283

Lowrie W, Channell J E T and Alvarez W 1980b A review of magnetic stratigraphy investigations in Cretaceous pelagic carbonate rocks *J. Geophys. Res.* **85** 3597

Lowrie W and Heller F 1982 Magnetic properties of marine limestones *Rev. Geophys. Space Phys.* **20** 171

Luterbacher H P and Premoli Silva I 1964 Biostratigrafia del limite Cretaceo–Terziario nelli Appennino Centrale *Riv. Ital. Palaeontol. Studigrafia* **70** 67

McDougall I 1977 The present status of the geomagnetic polarity timescale *The Earth: its origin, structure and evolution* ed M W McElhinny (New York: Academic)

McDougall I and Aziz-ur-Rahman 1972 Age of the Gauss–Matuyama boundary and the Kaena and Mammoth events *Earth Planet. Sci. Lett.* **14** 367

McDougall I and Chamalaun F H 1966 Geomagnetic polarity scale of time *Nature* **212** 1415

McDougall I, Saemundsson K, Johannesson H, Watkins N D and Kristjansson L 1977 Extension of the geomagnetic polarity timescale to 6.5 Myr: K–Ar dating, geological and palaeomagnetic study of a 3500 m lava succession in western Iceland *Bull. Geol. Soc. Am.* **88** 1

McDougall I and Watkins N D 1973 Age and duration of the Reunion geomagnetic polarity event *Earth Planet. Sci. Lett.* **19** 443

McDougall I, Watkins N D and Kristjansson L 1976a Geochronology and palaeomagnetism of a Miocene–Pliocene lava sequence at Bessastadaa, eastern Iceland *Am. J. Sci.* **276** 1078

McDougall I, Watkins N D, Walker G P L and Kristjansson L 1976b Potassium–argon and palaeomagnetic analysis of Iceland's lava flows: limits on the age of anomaly 5 *J. Geophys. Res.* **81** 1505

McKee E H and Elston D P 1980 Reversal chronology from a 7.9 to 11.5 Myr old volcanic sequence in Central Arizona: comparison with ocean floor polarity record *J. Geophys. Res.* **85** 327

Maenaka K 1979 Palaeomagnetic study of sediments around the Komyoike volcanic ash horizon in Osaka Group, Sempoku area, Osaka Prefecture, Japan *Geophys. Res. Lett.* **6** 257

Mankinen E A and Dalrymple G B 1979 Revised geomagnetic polarity timescale for the interval 0–5 Myr BP *J. Geophys. Res.* **84** 615

Mankinen E A, Donnelly J M and Grommé C S 1978 Geomagnetic polarity event recorded at 1.1 Myr BP on Cobb Mountain, Clear Lake volcanic field, California *Geology* **6** 653

Mankinen E A and Grommé C S 1982 Palaeomagnetic data from the Coso Range, California and current status of the Cobb Mountain normal geomagnetic polarity event *Geophys. Res. Lett.* **9** 1279

Napoleone G, Premoli Silva I, Heller F, Cheli P, Corezzi S and Fischer A G 1983 Eocene magnetic stratigraphy at Gubbio, Italy and its implications for Palaeocene geochronology *Geol. Soc. Am. Bull.* **94** 181

Ness G, Levi S and Couch R 1980 Marine magnetic anomaly timescales for the Cenozoic and Late Cretaceous; a précis, critique and synthesis *Rev. Geophys. Space Phys.* **18** 753

Ninkovich D, Opdyke N D, Heezen B C and Foster J H 1966 Palaeomagnetic stratigraphy, rates of deposition and tephra-chronology in North Pacific deep-sea sediments *Earth Planet. Sci. Lett.* **1** 476

Opdyke N D 1972 Palaeomagnetism of deep-sea cores *Rev. Geophys. Space Phys.* **10** 213

Opdyke N D, Burckle L H and Todd A 1974 The extension of magnetic timescale in sediments of the Central Pacific Ocean *Earth Planet. Sci. Lett.* **22** 300

Opdyke N D, Glass B, Hays J D and Foster J H 1966 Palaeomagnetic study of Antarctic deep-sea cores *Science* **154** 349

Opdyke N D, Lindsay E H, Johnson N M and Downs T 1977 The palaeomagnetism and magnetic polarity stratigraphy of the mammal-bearing section of Anza Borrego State Park, California *Quat. Res.* **7** 316

Picard M D 1964 Palaeomagnetic correlation of units within Chugwater (Triassic) Formation, West-Central Wyoming *Am. Assoc. Pet. Geol. Bull.* **48** 269

Premoli Silva I 1977 Upper Cretaceous–Palaeocene magnetic stratigraphy at Gubbio, Italy. II. Biostratigraphy *Geol. Soc. Am. Bull.* **88** 371

Rea D K and Blakely R J 1975 Short wavelength magnetic anomalies in a region of rapid seafloor spreading *Nature* **255** 126

Roggenthen W M and Napoleone G 1977 Upper Cretaceous–Palaeocene magnetic stratigraphy at Gubbio, Italy. IV. Upper Maastrichtian–Palaeocene magnetic stratigraphy *Geol. Soc. Am. Bull.* **88** 378

Rutten M C 1959 Palaeomagnetic reconnaissance of mid-Italian volcanoes *Geol. Mijnbouw.* **21** 373

Schlich R 1981 Echelle chronologique des inversions du champ magnétique terrestre pour L'Éocène, le Paléocène et le Crétacé Supérieur *Phys. Earth Planet. Int.* **24** 191

Shackleton N J and Opdyke N D 1973 Oxygen isotope and palaeomagnetic stratigraphy of equatorial Pacific core V28-238: oxygen isotope temperatures and ice volumes on a 10^5 and 10^6 year scale *Quat. Res.* **3** 39

Shuey R T, Brown F H and Croes M K 1974 Magnetostratigraphy of the Shungura Formation, south western Ethiopia: fine structure of the Lower Matuyama polarity epoch *Earth Planet. Sci. Lett.* **23** 249

Steiger R H and Jäger E 1977 Subcommission on geochronology: convention on the use of decay constants in geo- and cosmochronology *Earth Planet. Sci. Lett.* **36** 359

Steiner M B 1978 Magnetic polarity during Middle Jurassic as recorded in Summerville and Curtis Formations *Earth Planet. Sci. Lett.* **38** 331

Steiner M B, Shive P N and Shoemaker E M 1977 Polarity of the magnetic field during the upper and middle Jurassic *Trans. Am. Geophys. Union* **58** 376

Talwani M, Windisch C C and Langseth M G Jr 1971 Reykyanes ridge crest: a detailed geophysical study *J. Geophys. Res.* **76** 473

Valencio D A, Linares E and Villas J F 1970 On the age of the Matuyama–Gauss transition *Earth Planet. Sci. Lett.* **8** 179

Vandenberg J and Wonders A A H 1980 Palaeomagnetism of Late Mesozoic pelagic limestones from the southern Alps *J. Geophys. Res.* **85** 3623

Watkins N D 1968 Short-period geomagnetic polarity events in deep-sea sedimentary cores *Earth Planet. Sci. Lett.* **4** 341

——1972 Review of the development of the geomagnetic polarity timescale and discussion of prospects for its firm definition *Geol. Soc. Am. Bull.* **83** 551

——1976 Polarity subcommission sets up some guidelines *Geotimes* **21** 18

Watkins N D and Abdel-Monem A 1971 Detection of the Gilsa geomagnetic polarity event on the island of Madeira *Bull. Geol. Soc. Am.* **82** 191

Watkins N D, Kristjansson L and McDougall I 1975 A detailed palaeomagnetic survey of the type location for the Gilsa geomagnetic event *Earth Planet. Sci. Lett.* **27** 436

Watkins N D and Walker G P L 1977 Magnetostratigraphy of castern Iceland *Am. J. Sci.* **277** 513

7 The Earth's magnetic field and climate

7.1 Introduction

In this chapter possible relationships between the Earth's magnetic field and climate will be discussed. The field will be widened to include possible solar/weather relationships since the Earth's magnetic field could be indirectly involved. As will be seen this is a rather 'grey' area and, in spite of the increasing interest and vast literature on the subject, little convincing evidence for correlations has been obtained. Another grey area, possible relationships between magnetic reversals and faunal extinctions, will be discussed in §7.5. Possible correlations with other geophysical phenomena will not be discussed since, in general, they are even more speculative. As an example Glass and Heezen (1967) pointed out that the great field of tektites covering Australia, Indonesia and a large part of the Indian Ocean fell about 700 000 yr ago at about the time of the last magnetic reversal. They suggested that the fall of the body from which the tektites were formed killed the now extinct radiolaria and gave a jolt to the Earth, disturbing motions in the core and causing the dynamo to reverse. But as Bullard (1968) pointed out, it is 'difficult to believe that the fall of a large meteorite could selectively kill certain species of radiolaria all over the world and yet spare the kangaroos near the point of fall'.

Again, there appears to be no record of tektites or microtektites associated with the times of other field reversals (there were more than seventy during the last 20 Myr) and it is difficult to visualise how the impact of even quite a large mass (it has been estimated as about 100 million tons for the Australasian tektites) could cause a reversal of the Earth's magnetic field, whose origin lies within the core. Such a correlation is almost certainly due to chance and it is unlikely that there can be any physical connection between the two events.

Heirtzler (1970) speculated that increased earthquake activity and related increased upper mantle activity may be linked with magnetic changes. He suggested that a sufficiently large earthquake may cause a wobble of the Earth's spin axis and create a magnetic reversal. Kennett and Watkins (1970) extended this idea by suggesting that an increase in upper mantle activity should be manifested by increased volcanic activity, and thus evidence of such activity should be found at times of magnetic reversals. They presented evidence from late Cenozoic deep-sea sediment sequences from the Southern Ocean indicating increased amounts of volcanic ash close to the times of reversals. The theory was expanded by suggesting that if increased episodes of explosive volcanicity have occurred at specific times in the past, these possibly also affected the global climate and, in turn, faunal changes. Chappel (1975) has critically examined many of these suggestions and, on theoretical grounds, dismissed most of the proposed relationships.

The development of palaeomagnetic stratigraphy has played a vital role both in the dating and in the correlation of climatic history recorded in cores. The earliest palaeomagnetic studies on palaeoclimatic history were carried out, not on marine sediments, but on land-based sections in Iceland by McDougall and Wensink (1966). As a result of palaeomagnetic dating of basaltic lava flows containing embedded tillites, they were able to show that the onset of Icelandic glaciation occurred about 3 Myr ago, providing important information on the development of northern hemisphere glaciations during the late Cenozoic. The first attempt to use magnetostratigraphy for dating climatic history in marine sequences was made by Opdyke *et al* (1966), Hays and Opdyke (1967), and Hays (1967), all of whom carried out palaeomagnetic dating of icerafted debris in Antarctic piston cores. Later Goodell *et al* (1968) and Goodell and Watkins (1968) obtained important information on the history of Antarctic glaciation based on the palaeomagnetic dating of glacial–marine sequences collected during the USNS Eltanin expeditions. Of particular importance was their demonstration that sediment rafting by icebergs began to occur in the Southern Ocean earlier than 5 Myr BP (their oldest dated core) so that major glaciation developed much earlier in the Antarctic than in the northern hemisphere. They also showed that the Antarctic ice sheet has been a permanent feature for at least the last 5 Myr.

7.2 The orbital climatic theory of Milankovitch

The first suggestion that orbital variations might affect climate was made in 1830 by John Herschel. The idea was revived many times, more recently by Milankovitch in 1941 with whose name it is now associated. The theory holds that regular, predictable changes in the orientation of the Earth's axis of rotation and the shape of its orbit affect the distribution of sunlight over the Earth's surface. The tilt of the Earth's axis away from the plane of its orbit (its obliquity ε) varies between 22.1° and 24.5° (it is now 23.5°). Since it is the obliquity which causes seasons, a cyclic variation in the obliquity will in turn produce a cycle in the strength of the contrast between seasons. The Earth's axis of rotation also precesses, describing a small circle among the stars. This is measured as the angular distance ω of perihelion from the autumnal equinox. Precession also affects the contrast of the seasons since it determines at what point in the Earth's orbit winter and summer occur. Winters occurring near the Earth's closest approach to the Sun would be warmer on average than those occurring at its farthest point. Finally the eccentricity e of the Earth's orbit, which is a measure of the Sun's departure from the geometric centre of the Earth's orbit, is not constant, varying between 0 and 0.06. Integrated over all latitudes and over an entire year, the energy influx depends only on e. However, the geographic and seasonal pattern of irradiation essentially depends only on ε and on $e \sin \omega$.

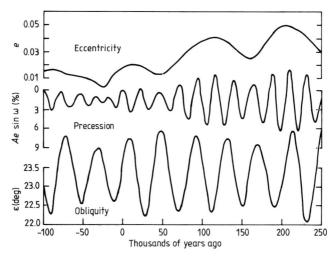

Figure 7.1   Variations in orbital geometry as a function of time (after Imbrie and Imbrie 1980).

Each of the orbital elements is a quasi-periodic function of time (see figure 7.1). Although the curves have a large number of sinusoidal components, calculated spectra are dominated by a small number of peaks. The most important term in the series expansion for eccentricity has a period calculated by Berger (1977b) to be 413 000 yr. Eight of the next twelve terms range from 95 000 to 136 000 yr. In low-resolution spectra, these terms contribute to a peak that is often loosely referred to as the 100 000 yr eccentricity cycle. In contrast, the spectrum of the obliquity is relatively simple, being dominated by components with periods near 41 000 yr. The main components of precession have periods near 23 000 and 19 000 yr. In low-resolution spectra, these are seen as a single peak near 22 000 yr.

Over the past 600 000 yr, almost all climatic records are dominated by variance components in a narrow frequency band centred near a 100 000 yr cycle. Yet a climatic response at these frequencies is not predicted by the

Milankovitch theory—or any other astronomical theory that involves a linear response. It should be stressed that neither the cyclic variation in obliquity nor precession involve a change in the total amount of sunlight insolation falling on the Earth—they merely affect how much sunlight a particular latitude receives at a particular season.

There have been many possible explanations put forward to explain the cause of fluctuations in the Pleistocene ice sheets. However only the orbital theory of Milankovitch (1941) can be tested geologically since it is the only theory that predicts the frequencies of the major fluctuations. There are two difficulties in carrying out such a test—the uncertainty in identifying which aspects of the radiation budget are critical to climatic change, and the uncertainty of geological chronology. Hays *et al* (1976) attempted to test the hypothesis by considering secular changes in the orbit as a forcing function of a system whose output is the geological record of climate, without identifying, or evaluating, the mechanism through which climate is modified by changes in the global pattern of incoming radiation. Most of their climatic analysis is based on the assumption that the climatic system responds linearly to orbital forcing. They used two cores from the southern Indian Ocean and for geological data for the last 450 000 yr chose the $\delta^{18}O$ composition of planktonic foraminifera, an estimate of summer sea-surface temperature at the core site derived from a statistical analysis of radiolarian assemblages and the percentage of *Cycladophora davisiana*, the relative abundance of a radiolarian species not used in the estimation of sea-surface temperatures (see figure 7.2). They carried out both frequency-domain and time-domain tests. Over the frequency range 10^{-4}–10^{-5} cycles/yr, they found that the climatic variance of these three measures of global climate is concentrated in three discrete spectral peaks at periods of 23 000, 42 000 and 100 000 yr (see figure 7.3), which correspond to the dominant periods of the Earth's solar orbit (precession, obliquity and eccentricity). Moreover the ratio of the two frequencies (obliquity and precession) detected in the cores does not differ significantly from the predicted ratio (~ 1.8). In the time-domain analysis they used two filters, one centred at a frequency of 0.25 cycles/1000 yr (the 40 K filter) and one at 0.043 cycles/1000 yr (the 23 K filter). They found that over the past 300 000 yr each of the 40 K components of the geological record showed a constant phase relationship with obliquity and the 23 K components a constant phase relationship with precession. The dominant 100 000 yr climatic component has an average close to, and in phase with, orbital eccentricity. Hays *et al* suggested that such a correlation probably requires a non-linear response of the climatic system to orbital forcing. They concluded that changes in the Earth's orbital geometry are the fundamental cause of the succession of Quaternary ice ages.

Rooth *et al* (1978) have argued that power spectrum analysis of the oxygen isotope record cannot be used to search for evidence of astronomical forcing because parts of this record have a step-like character reflecting the rapidity of

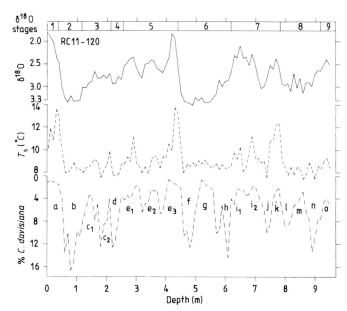

Figure 7.2   Depth plots of three parameters measured in core RC11-120: $\delta^{18}O$ (full curve). T_s (broken curve), and percentage of *C. davisiana* (chain curve). Letter designations of peaks on the latter curve are informal designations of various parts of the record (after Hays *et al* 1976).

major deglaciations. Implicit in their argument is the idea that the presence of second and higher order harmonics in a Fourier analysis of a signal of this type might introduce sufficient spectral energy to mask orbital influences operating over a frequency range from 1 cycle/19 000 yr to 1 cycle/41 000 yr. However Hays *et al*, in footnote 66 of their 1976 paper, considered this problem for three independent climatic time series and concluded that the observed spectral peaks cannot be explained as harmonic contaminants. Moreover, they presented evidence for systematic phase relationships with obliquity and precession rather than with the low-frequency components of the record. This rules out the possibility that the observed peaks in climatic spectra could originate in non-sinusoidal low-frequency components.

Weertman (1976) estimated the fluctuations in the size of ice age ice sheets using glacier mechanics and Milankovitch orbital variations, and concluded that solar radiation variations seemed to be large enough to account for ice ages. Milankovitch variations gave results similar to that produced by shifting the land surface north and south by about 1000 km. Suarez and Held (1976) made a preliminary attempt to test the Milankovitch theory using a zonally symmetric energy balance climate model forced with seasonally varying insolation. They found large discrepancies between the model and estimates of summer sea-surface temperature inferred from faunal abundances of plank-

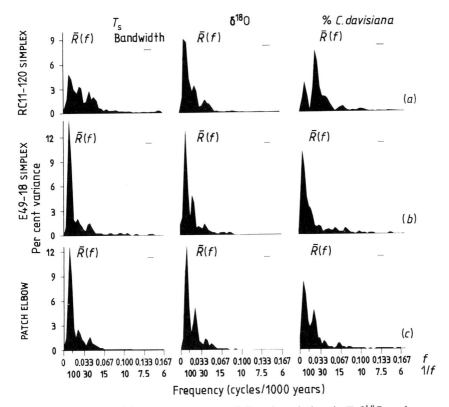

Figure 7.3 High resolution spectra of climatic variations in T_s, $\delta^{18}O$, and percentage of *C. davisiana*. Variance (as percentage of total variance per unit frequency band) is plotted as a function of frequency (cycles per thousand years). (*a*) Spectra for core RC11-120 calculated for the SIMPLEX age model. (*b*) Spectra for core E49-18 calculated for the SIMPLEX age model. (*c*) Spectra of the combined (PATCH) record calculated for the ELBOW age model. SIMPLEX and ELBOW are two chronological models and PATCH is a time series in which records from two cores have been joined — for details see Hays *et al* (1976).

tonic foraminifera in a single deep-sea core from the North Atlantic. In spite of those shortcomings, however, they suggested that a substantial portion of climatic variations can be explained as the response to perturbations in the orbital parameters.

Wigley (1976) has re-examined the results of Hays *et al* (1976). He pointed out that their orbital input spectra show a split peak near the precession period of 21 000 yr (table 4 of Hays *et al* gives peaks at periods of 23 100 and 18 800 yr). Wigley showed that non-linear interaction between these two frequencies could be expected to give an output signal with a period 101 000 yr. In addition to this effect, non-linear response to amplitude modulation of the precession signal will produce a spectral peak at the frequency of amplitude modulation: that is, at the eccentricity period of approximately 100 000 yr. These two mechanisms both lead to strong output power near a period of 100 000 yr with little or no input power at this period. The data of Hays *et al* also show significant power at periods corresponding to precession and obliquity so that the climate response function cannot be wholly non-linear. Wigley (1976) suggested that the response function has both linear and non-linear parts of approximately equal importance.

Berger (1977a) pointed out that the periods found by Hays *et al* (1976) are very close to those predicted by himself (1976, 1977b) in later, more accurate calculations of the variations of the various 'Milankovitch' parameters. The roughly 24 000 and 19 500 yr periods from the core samples are not significantly different from the periods associated with the largest amplitude terms in the series expansion of the precessional parameter $e \sin \omega$; the periods around 42 000 yr are essentially identical with the most important term in the expansion of the obliquity ε; and the peaks around 106 000 yr, containing most of the variance, might be regarded as either a contribution from the eccentricity, where a weighted mean of the main amplitude terms has a period of 110 753 yr or as a beat effect of precessional periods, as suggested by Wigley (1976). Berger (1976) also pointed out that the most important term in the series expansion of the eccentricity has a periodicity of 412 085 yr—too long to be seen in the analysis of Hays *et al*—and the convergence of the series is very slow. If longer cores can be obtained, Berger (1977a) forecast that the 100 000 yr peak would be split into two peaks at 95 000 and 123 000 yr with the main period of 413 000 yr being clearly evident.

Evans and Freeland (1977) criticised the analysis of Hays *et al* (1976) and do not believe that any cause and effect relationship between variations in the Earth's orbital parameters and changes in climate has been established. Their objections are both of the approach used and technical. They pointed out that the idea that the existence of peaks in the spectra of the geological series is indicative of forcing at those frequencies is a gross oversimplification of the concept of a time-invariant linear system. Such peaks are a consequence of the dynamics of the system, not just the spectral character of the input. They suggested that a far more convincing approach would be to assume astronomical forcing, compute the transfer function, and attempt to provide a physical model to account for the observed peaks. In reply Hays *et al* (1977) pointed out that the disagreements stem from fundamentally different strategies for studying a natural system which is complex and poorly understood. Their strategy is to make quantitative predictions based on the simplest and most general assumption—that the system is linear and time-invariant—and then to test both the orbital hypothesis and that assumption by searching for the predicted frequencies in three climatic spectra. The strategy of Evans and Freeland has many disadvantages and their alternative approach many problems. Hays *et al* reaffirmed their belief in their results, stressing that the main conclusion of their analysis is that orbital variations control the *timing* but not the amplitudes of the main ice age successions.

Another fundamental question is whether the three observed cycles account for most of the climate variability having periods in the range predicted by Milankovitch, or only for a small part of it. Hays *et al* (1976) originally suggested that about 80% of the total variability seen in their sediment record was associated with the three most obvious cycles—other workers have suggested it may be closer to 10% than 80%. A significant background of variability probably exists on which the Milankovitch cycles would be superimposed. The broad spectrum of background climatic variability commonly observed in geological records, especially the shorter period variability, could be generated by other processes, such as those in a stochastic, or random, theory of climatic variation. In this connection Kominz and Pisias (1979) examined a number of simple linear models of the record of the Earth's ice volume over the last 730 000 yr and concluded that, although forcing by variations in the Earth's orbital parameters of tilt and precession is real, it is less than 25% of the variation seen in global ice volume. They also found no evidence for a linear relationship between eccentricity and global climates and believe that Pleistocene glacial variations are largely stochastic in nature.

Kominz *et al* (1979) also devised a new $\delta^{18}O$ timescale based on a constant accumulation rate of aluminium which they showed to be valid for the past 360 000 yr by comparison with ^{230}Th ages. To facilitate comparison with Shackleton and Opdyke's (1973) timescale, they used the same core (V28-238) from the equatorial Pacific. Spectral and cross-spectral analysis of the $\delta^{18}O$ records of V28-238 and a detailed composite Indian Ocean record confirmed the conclusion of Hays *et al* (1976) that changes in the Pleistocene climate are a result of forcing by periodic fluctuations of the Earth's obliquity and precession. Kominz *et al* obtained a new timescale by tuning[†] the $\delta^{18}O$ record of core V28-238 to the record of the Earth's obliquity. Their aluminium timescale gives an age of 0.693 Myr for the Brunhes–Matuyama reversal in close agreement with its earlier estimate. Their second timescale gives a value of 0.728 Myr in excellent agreement with the revised estimate of 0.73 Myr (Mankinen and Dalrymple 1979).

Another problem is that most published climatic records that are more than 600 000 yr old do not exhibit a strong 100 000 yr cycle[‡], e.g. it is virtually absent

† By 'tuning' a model one means adjusting the model parameters over a range of physically reasonable values until the optimum model variant is identified.
‡ Calculations have often used orbital parameters calculated by Vernekar (1972). When they are compared with those using the more recent values of Berger (1977b, 1978) the timing of maxima and minima do not differ by more than 1000 yr over the past 400 000 yr. For older intervals, however, discrepancies between the calculations become significant, and the work of Berger is to be preferred because it includes the effect of more terms.

from isotopic records of planetary ice volume from earlier parts of the Pleistocene (Shackleton and Opdyke 1976). The problem of explaining the 100 000 yr cycle with a simple non-linear mechanism of the type suggested by Hays *et al* (1976) or Wigley (1976) poses a second problem. It is difficult to introduce substantial 100 000 yr power into the response without also introducing power reflecting the 413 000 yr eccentricity cycle in amounts that are much greater than have been detected in most climatic records.

Imbrie and Imbrie (1980) have argued for a fundamental change in research strategy: instead of using numerical models of climate to test the astronomical theory, one should use the geological record as a criterion against which to judge the performance of physically motivated models of climate. They further pointed out that even an excellent correlation between climate and a particular insolation curve $Q(t)$ is no assurance that physical mechanisms operating at the latitude and season represented by $Q(t)$ actually dominate the climatic response. This ambiguity exists because a curve nearly identical to $Q(t)$ may be expressed as a linear combination of curves for other latitudes and seasons.

Imbrie and Imbrie (1980) fitted simple non-linear mathematical models to $\delta^{18}O$ curves. They found that reasonable fits were obtained if the timescale for ice sheet growth is about 27 000 yr and for decay about 7000 yr. Oerlemans (1980) considered the problem of the 100 000 yr cycle in a similar way. Experiments with a northern hemisphere ice sheet model showed that the 100 000 yr cycle and its sawtooth shape may be explained by ice sheet/bedrock dynamics alone as first suggested by Weertman (1961). This means that the occurrence of a 100 000 yr peak in the power spectrum of the $\delta^{18}O$ record is unrelated to variations in eccentricity. Recent work by Kominz and Pisias (1979) strongly supports this view: they found that coherence spectra of an oxygen isotope record and the orbital parameters showed no significant peak around a period of 100 000 yr. Oerlemans' work may however support the Milankovitch theory. The strength of the 100 000 yr periodicity is not constant and almost vanishes about 1 Myr BP. This would be difficult to understand if the cause were purely internal, but less so if climatic changes were the result of orbital forcing. The lack of a 100 000 yr periodicity in the early climate record may be related to the fact that perennial sea-ice cover in the Arctic only began around 700 000 yr ago (Margolis and Herman 1980). However, it may well be that around 1 Myr BP, the orbital parameters themselves, if more accurately known, have no 100 000 yr signal.

Kukla *et al* (1981) noted that a specific orbital configuration (high obliquity combined with the June perihelion) marked the beginning of the past three interglacials, from which they inferred that the primary cause of the glacial cycle may be astronomical. They thus introduced a purely empirical astronomical climate index (ACLIN) combining the three orbital variables to predict the major climate changes in the late and middle Pleistocene and the near future. ACLIN closely correlates with the major climatic events revealed by independently dated proxy climate indicators of the past 130 000 yr—it also successfully differentiates the interglacials and shows a 100 000 yr periodicity.

A record of orbital variations preserved in the sediments at the Deep Sea Drilling Program's (DSDP) site 158 in the equatorial Pacific has been found by Moore *et al* (1982). The orbital cycles appear as variations in the amount of calcium carbonate in sediments deposited between 5 and 8.5 Myr ago. Moore *et al* also studied the carbonate cycles in sedimentary cores deposited over the past 2 Myr. They identified three different cycles of changing carbonate content, one having a period of about 400 000 yr. Using cross-spectral analysis, they demonstrated a correlation between the eccentricity cycle and the carbonate content of the cores, showing that the two were in phase over the last 8 Myr. Two other orbital cycles, which had already been identified in sediments of the past half million years also showed up in the 2 Myr long Pleistocene carbonate record. One cycle had a period of 100 000 yr and was in phase with a 100 000 yr cycle in eccentricity. A minor cycle of 41 000 yr matched the variations in the tilt of Earth's axis.

Although the 100 000 yr cycle dominated climatic variability during the Pleistocene ice ages, accounting for more than half of the variability due to the three orbital cycles and 29% of the total variability, it had a minor effect 5–8 Myr ago (see figure 7.4). Even after the 400 000 Myr cycle was used to adjust

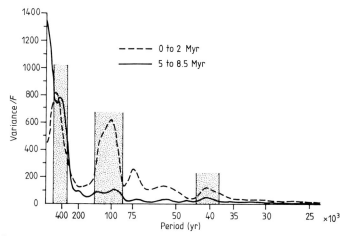

Figure 7.4 Variance spectra showing how much stronger the effects of the 100 000 yr cycle were over the past 2 Myr (broken curve) than 5 to 8 Myr ago (full curve). The larger the peak, the more variability in the carbonate content of the sediment (and presumably in climate) is attributable to cycles having that period. The variability attributable to the 400 000 yr cycle remained the same, but that in the 100 000 yr cycle was about six times greater during the more recent period. Variability having periods between 40 000 and 100 000 yr also increased dramatically (after Moore *et al* 1982).

or 'tune' the dating of the core, the 100 000 yr cycle accounted for six times less variability than it did during the Pleistocene. The variability of the 400 000 yr cycle remained unchanged, which means that the variability of the carbonate record, and presumably of the climate, was twice as great during the past 2 Myr as it was 7 Myr ago.

7.3 The Earth's magnetic field and climate

Wollin *et al* (1971a, b) examined deep-sea sedimentary cores for the past 1.2 Myr and claimed to have found a correlation between the intensity of magnetisation and climatic changes as represented by variations in the sediments in the abundance of the planktonic foraminifera *Globorotalia menardii* and the coiling direction of *Globorotalia truncatulinoides*. They concluded that there was a relationship between changes in the Earth's magnetic field and climate—higher magnetic intensity producing or being produced by colder climate. Amerigian (1974) pointed out, however, that there are at least equally feasible causes of the observed correlations found by Wollin *et al*. One of the principal factors affecting the intensity of magnetisation of sediments is the intensity of magnetisation of the particles which settle to form the sediments, and this in turn depends on composition. Most of the cores examined by Wollin *et al* were high in organic calcium carbonate and Amerigian showed that the greater the proportion of essentially non-magnetic calcium carbonate present in a sediment the lower (statistically) is the sediment's intensity of magnetisation. Thus, calcium carbonate production, which is climatically controlled, can affect the intensity of magnetisation in a way which is completely unrelated to the strength of the geomagnetic field.

Kent (1983) is also sceptical about the correlation between the Earth's magnetic field and climatic changes claimed by Wollin *et al* (1971b) from a study of the record of deep-sea core V20-108. He points out that the original inclination data, published earlier by Ninkovich *et al* (1966), show some important discrepancies with later versions of the same data. The original data show that below 300 cm the inclinations fell close to the expected dipole value of 63.5° (positive for normal and negative for reversed polarity). In contrast, anomalous values of the inclination are found in the upper 300 cm, being generally shallower than the dipole value and recording intervals of negative inclination during the Brunhes normal epoch. Kent believes that the magnetic data from the upper 300 cm of core V20-108 do not represent geomagnetic field changes. In two nearby cores—V20-107 taken 227 km to the south and V20-109 taken only 208 km to the north of V20-108—the magnetic data are of high internal consistency and no such anomalous inclinations are observed in the Brunhes (Ninkovich *et al* 1966). Such localised occurrences of large departures from an axial dipole field direction are difficult to explain with sources in the Earth's core and are more likely to be due to a distorted record.

Kent also points out that examination of the Lamont coring and curatorial logs for V20-108 reveals that at least the upper 300 cm of the sediment was drawn through a bent pipe during coring, a condition that could easily lead to disturbance of the sediment and the magnetic record. The same logs also show that the sediment core was broken in several places on board ship, in particular at 298 cm to facilitate transport and at 460 cm where coring pipes were uncoupled; single-sample measurements of anomalous magnetic inclination closely correspond to these levels.

More recently Wollin *et al* (1977, 1978) added a third factor—the eccentricity of the Earth's orbit. From an analysis of deep-sea sediment cores they claimed that, over the past 2 Myr, high eccentricity of orbit has corresponded with low magnetic field intensity and warm climate, and they concluded tentatively that the orbital eccentricity partially modulates both the magnetic field and climate. Figure 7.5 shows changes in climate (as indicated by faunal variations) and variations in geomagnetic intensity in two deep-sea sediment cores compared with fluctuations in the eccentricity of the Earth's orbit. Chave and Denham (1979) have argued that no connection between climate and magnetic intensity has been established, the correlation found by Wollin *et al* being due to their failure to remove the effect of the varying input of magnetic material which is largely climatically controlled. It is easy then to see why the ancient magnetic field might correlate with climate. Ruddiman and McIntyre (1976) showed that the flows of biogenic and terrigenous matter (i.e. material which contains magnetic constituents) into oceanic sediments vary by factors of up to at least 3 depending upon climatic conditions. During glacial periods the relative contribution of terrigenous material increases. Generally then the NRM of sediments produced during glacial times should be higher than that of interglacial sediments even if the strength of the geomagnetic field were to remain constant, and whatever the variations in the Earth's eccentricity. Chave and Denham supported their contention by estimating the variation of palaeointensity along a core from the North Atlantic deposited between 127 000 and 60 000 yr ago. They used, not raw NRM values, but values of NRM normalised against an imposed anhysteretic remanent magnetisation, ARM. Whereas both NRM and ARM vary with the quantity of magnetic material present, the ratio of the two should not. The palaeointensity they obtained showed no correlation at all with climate as represented by faunal variations throughout the core.

Harrison (1974) had earlier maintained that those cores which show correlations between direction and/or intensity of magnetisation and climatic indicators are not accurately recording the relevant parameters of the Earth's magnetic field—the correlation being caused by climatic effects which have a direct influence on the magnetisation of the sediments. For example, Bonatti and Gartner (1973) showed, in a core from the Caribbean, that there are several geochemical changes produced directly by climatic effects which could easily affect the authogenic production of magnetic material in such a way as to

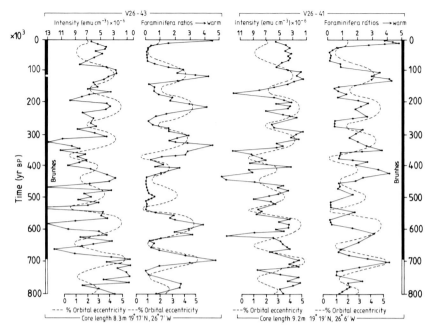

Figure 7.5 Comparison of variations in geomagnetic intensity, climatic changes, and fluctuations of the eccentricity of the Earth's orbit. The deep-sea sediment cores V26-43 and V26-41 are from the north-east Atlantic. The climatic curves are based on variations in abundance of the *Globorotalia menardii* group of foraminifera according to the method developed by Ericson and Wollin (1956). The timescale for the climate and intensity curves is based on the boundary condition of 0.69 Myr for the Matuyama–Brunhes boundary, zero age for the tops of the cores, and an assumption of constant sedimentation rate. The orbital eccentricity variations (broken curves) are after Van Woerkom (1953) (after Wollin *et al* 1977).

cause the correlation between intensity and climate claimed by Wollin *et al* (1971a).

K ent (1982) has also cautioned against accepting the apparent correlation of palaeomagnetic intensity with climatic records in deep-sea sediments. He showed a pronounced dependence of the NRM intensity on sediment composition. The observed correlation between NRM intensity and climate (e.g. the $\delta^{18}O$ record) can be largely accounted for as the result of an intermediary lithological effect. Decreased carbonate content during glacials results in increased concentrations of magnetic material which in turn contribute to higher NRM intensities. Kent also found cases where high NRM intensities characterised sediments deposited during glacial intervals in one region, but were associated with sediments deposited during interglacials in another.

Wollin *et al* (1973) also considered the possibility of a correlation between short period changes in the Earth's magnetic field and climate. They tentatively concluded that the trends in intensity from most magnetic observatories over the last 30 yr correlate negatively with the 10 yr means of air temperature. However the correlations were made on a geographical basis, and any attempt to seek a relationship between solar activity and meteorological phenomena should look for global responses rather than localised, isolated effects. Sternberg and Damon (1979) have also criticised Wollin *et al*'s (1973) suggestion that trends in the intensity of most magnetic observatories with records over at least 30 yr correlate negatively with 10 yr means of air temperature. They pointed out that data from only 22 of the 43 magnetic observatories listed were presented, that the use of 10 yr temperature means yields only three or four data points and finally that the degree of correlation is not quantitatively evaluated.

Sternberg and Damon give statistical evidence to show that there is no globally inverse relationship between geomagnetic intensity and climate for the period covered by observatory records. Since geomagnetic intensity and climate are both regional parameters showing temporal and spatial coherence, data used to analyse the hypothesis should be well distributed geographically—otherwise a spurious regional correlation might bias the overall results. Because of the highly linear trends in the variation of magnetic field strength in many parts of the world, there will inevitably be some places where there will be an inverse relationship between changes in temperature and the magnetic field. Without a viable physical mechanism that could produce such a relationship worldwide, Sternberg and Damon maintain that the weak, negative correlation between magnetic field strength and temperature in the historical record is a fortuitous result of the linear secular variation of magnetic field intensity and the poor geographic distribution of stations.

Roberts and Olson (1973a, b) studied days on which geomagnetic activity showed a sizable increase, which was assumed to have a solar cause. They also studied the history of cyclones (low-pressure troughs) from the Gulf of Alaska as they moved across the continental United States, and found that troughs associated with geomagnetic activity were significantly larger on the average than troughs associated with intervals of quiet geomagnetic conditions. These results relate a vorticity area index, a measure of the size of low-pressure troughs devised by Roberts and Olson and which can be computed from maps of the height of constant pressure (300 mbar) surfaces, to times of passage of the solar magnetic sector boundaries which sweep across the Earth. Knight and Sturrock (1976) have carried out a spectral analysis to see if there is any correlation between solar activity and weather. They used as data the geomagnetic activity index A_p and the vorticity area index and concluded that their analysis, although not establishing a connection between these two parameters, does provide some supporting evidence for such a connection. Stolov and Shapiro (1974) also found strong statistical evidence for a relationship between solar magnetic storms and the subsequent behaviour of the 700 mbar contour height. This revises their earlier (1971) conclusion. The reason for their change of opinion is that ordinary meteorological factors introduce a strong seasonal trend in 700 mbar heights and this must be removed before the solar geomagnetic effect can be seen.

King (1974a) noted a similarity between the pattern of the intensity of the Earth's magnetic field over the northern hemisphere and the mean height contours of surfaces of constant atmospheric pressure—both having an elongated, dumb-bell like form. He suggested that the Earth's magnetic field may possibly influence the average tropospheric pressure system (500 mbar) at high latitudes. If this were true, he then argued that changes which occur in the pattern of 'permanent' depressions in the troposphere as the magnetic field varies may be accompanied by climatic changes. He further noted that the average pressure system seems to move westward as the non-dipole component of the magnetic field drifts westward, although not at the same rate. Chiu (1974) attempted to test King's (1974a) suggestions by correlating archaeomagnetic and archaeoclimatic observations from various parts of the world. He found only a tenuous correlation at best. He made the further suggestion that perhaps the climatic regime has a global component together with a drifting component. Sawyer (1974) also raised objections to King's hypothesis, claiming amongst other things that there has been some selection of data—the associations claimed by King being much better for 60°N in the winter than for other latitudes and seasons. Although King (1974b) replied to some of Sawyer's criticisms he ignored the main criticism. The numerical models of the general circulation of the atmosphere predict wave number 3 at lower latitudes (around 40°N), as observed, whereas the geomagnetic field does not, and cannot, have wave number 3 in it. King later (1975) reasserted his claim of a correlation between the longitudinal variations of the height of the 500 mbar level (shifted an arbitrary 25° to the west) and geomagnetic intensity at 60°N. However, as Pittock (1978) has pointed out, two phenomena which both exhibited wave number 2 spatial patterns at this latitude (one probably because of topographic effects, the other because it stems from a magnetic dipole) must, with a suitable arbitrary phase shift, give a very high spatial correlation.

Bucha (1976a, b, 1977) has long advocated a connection between geomagnetism and the Earth's climate and proposed a model (1980) which links the position of the geomagnetic pole with the causes of glacial and interglacial periods. He suggested that changes in the intensity of corpuscular radiation (as indicated by geomagnetic activity) significantly affect the temperature (see figure 7.6) and pressure patterns over the geomagnetic pole and polar regions causing a pronounced modification of the general atmospheric circulation. Bucha maintains that the recurrence of significant warming and cooling does not take place simultaneously over the Earth as a whole. Thus when glacial periods occurred in Europe and the North Atlantic, interglacial periods were

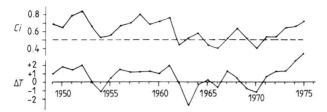

Figure 7.6 Correlation of geomagnetic activity (Ci-indices) and deviations of the temperature from the standard in Prague for the months of November–February (mean values) from 1949 to 1975 (after Bucha 1980).

present in the Pacific and vice versa in step with changes in the position of the geomagnetic pole from the Atlantic to the Pacific hemisphere.

Lambeck and Cazenave (1976) obtained a correlation coefficient of 0.85 between variations in the length of the day (LOD) and average temperatures over the whole Earth. Courtillot *et al* (1978) later obtained a correlation between secular variation accelerations and extrema in LOD fluctuations. This was put on a firmer basis by Le Mouel *et al* (1981) using records from European observatories going back to 1685—the correlation coefficient being approximately 0.8 (see figure 7.7). Courtillot *et al* (1982) suggested that there may be a possible long-term influence of core motions on climate—angular momentum being transferred first from the core to the mantle and finally to the atmosphere.

Pittock (1978) has written a long critical review on Sun–weather relationships. The literature on the subject is enormous—a bibliography by Shapley *et al* (1975) covering the years 1958–1975, with selected coverage before 1958, contains more than 800 references. Despite this great number of papers, Pittock concluded that little convincing evidence has yet been produced for any real correlations between sunspot cycles and climate on the 11 and 22 year timescales. He showed that the literature is full of negative findings and contradictions even when the latest techniques are applied to the most comprehensive regional and global meteorological data sets. Most of the climatically imported atmospheric and weather variables (e.g. temperature, precipitation or ozone content) show day-to-day, seasonal and year-to-year variations which are usually comparable to, or larger than, variations in longer term mean values. Even if more data and better analysis enable the detection of statistically significant relationships, they are likely to account for so little of the total variation in the climatic record as to be of little practical value. Pittock concedes however that there appears to be some evidence for significant correlations on a timescale of days between certain meteorological indices and the passage of the solar magnetic sector structure across the Earth (see e.g. Wilcox 1975, Hines and Halevy 1977).

The main problem in establishing a possible connection between variable

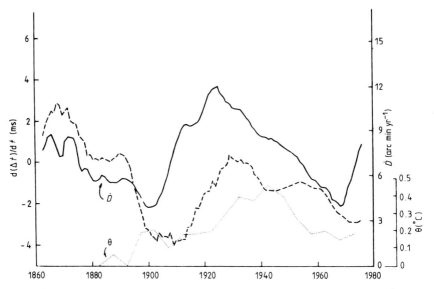

Figure 7.7   Plot of the secular variation of the geomagnetic declination $\dot{D}$ (full curve), of the excess length of the day $d(\Delta t)/dt$ (broken curve), and of successive 5 yr coverages of the temperature θ over the whole Earth, expressed as departures from the means for 1880–84 (dotted curve) (after Courtillot *et al* 1982).

conditions on the Sun and weather and climate is that of identifying a likely physical link between the Sun and the lower atmosphere. Possible solar characteristics that might be responsible are changes in the solar constant, sunspot numbers (and dimensions), solar cosmic rays, galactic cosmic rays (as modulated by solar activity), sector boundary crossings, ultraviolet emissions and geomagnetic effects. Observations and calculations of the effect of variations in the solar constant have been inconclusive. Analyses of spacecraft measurements have recently established that the total radiative output of the Sun varies at the 0.1–0.3% level (Eddy *et al* 1982). The fluctuations may be modelled as radiative deficits proportional to the area of the solar disc covered by sunspots. Eddy *et al* estimated that climatic responses are insignificant compared with normally occurring changes—changes in the solar constant perturbing terrestrial surface temperatures by at most a fraction of a degree. Again the combined energy of the solar wind, cosmic rays and extreme ultraviolet radiation is less than one-millionth of the energy of the visible radiation. Ultraviolet energy is one-tenth, but most of this energy never reaches the ground, being deflected or absorbed before it penetrates the upper atmosphere. Changes in these forms of solar energy do produce noticeable changes in the rarefied upper atmosphere, but cannot directly influence the dense lower atmosphere. What is needed is some mechanism that could trigger changes in the lower atmosphere. The more specific question of a possible correlation between climate and reversals of the Earth's magnetic field will be considered in the next section.

7.4 Magnetic reversals and changes in climate

Kennett *et al* (1971) suggested that there may be a relationship between palaeoclimatic and geomagnetic polarity changes. Cooling trends commonly began at, or shortly after, geomagnetic polarity changes (e.g. at the top of Kaena, base of Olduvai, upper part of Olduvai, and base of Gilsa) or cooling maxima coincided with geomagnetic reversals (e.g. at the Matuyama–Gauss boundary). Exceptions are the Mammoth and base of Kaena, which do not coincide with any climatic changes, and the top of the lower part of the Olduvai, which occurred immediately before a warming trend. It must not be forgotten, however, that the dating of these magnetic events is not precise and some of the magnetic boundary dates have changed (see §6.3).

Roberts and Olson (1973a) suggested that high latitude atmospheric ionisation plays an important role in nucleating cirrus clouds. If this is true, Harrison and Prospero (1974) argued that a large reduction in the strength of the Earth's magnetic field (as would be the case during a reversal) would lead to increased upper atmospheric cloudiness at all latitudes and major changes in climate—warming at high latitudes and considerable cooling in middle latitudes. Fairbridge (1977) attempted to test this suggestion by looking for possible climatic changes that may have occurred at the time of the so-called Gothenburg excursion (figure 7.8), dated by Morner and Lanser (1975) as occurring between 13 750 and 12 350 yr BP (see §4.5). Figure 7.8 shows that around the time of the Gothenburg excursion there was a short readvance of glaciation during the general melting that was taking place during the retreat of the great Laurentian ice sheet. There was also a dramatic fall in sea level at this time of general rise, and a very greatly increased discharge from the river Nile. Hecht (1977) has criticised Fairbridge's correlation, pointing out that he ignores the cyclical nature of continental and oceanic climatic histories. The Holocene is characterised by alternating intervals of glacial advance and retreat and Hecht maintains that one cannot justify correlating the Gothenburg excursion with just a small segment of climatic history. He also argued that the δ^{18}O record of deep-sea cores shows no major climatic changes at the time of the Gothenburg excursion.

Doake (1977) has suggested that there is a connection between ice ages and reversals of the Earth's magnetic field. The moment of inertia of the Earth will change by variations in the size of polar ice sheets and the resulting redistribution of water masses. To conserve angular momentum, the Earth's rate of rotation must also change. Doake thus suggested that the generation mechanism of the Earth's magnetic field may be affected by changing conditions at the MCB. He carried out simple order of magnitude calculations

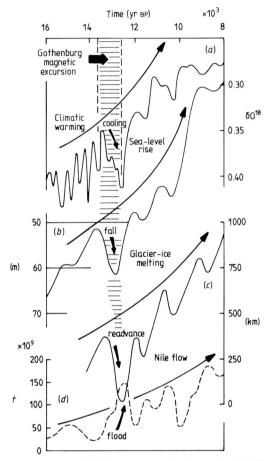

Figure 7.8   Four climatic indicator curves for the period 16 000 to 8000 yr
BP. The Gothenburg palaeomagnetic excursion occurred between 13 750–
12 350 yr BP (Morner and Lanser 1975). (*a*) Climatic warming reflected by
$^{18}O/^{16}O$ ice records from Camp Century, Greenland; (*b*) sea-level rise,
from data assembled by Fairbridge; (*c*) glacier-ice melting, measured as
km of retreat in southern Canada; (*d*) Nile discharge, measured from the
sedimentary cut and fill levels in Nubia (Fairbridge 1962) (after Fairbridge
1977).

which indicated that a continuous change in the speed of rotation over a
sufficiently long time could alter conditions at the MCB sufficiently to perturb
the magnetic field and perhaps cause it to reverse its polarity. It is doubtful
however if the effect is strong enough. Moreover, although such a result seems
plausible, conclusive evidence for a correlation between glacial periods and
changes in the magnetic field is hard to obtain because of the difficulty in
identifying and accurately dating variations in the two parameters.

In a later paper, Doake (1978) analysed statistically the probability of a correlation between climatic changes and field reversals. He compared dates of climatic episodes as recorded in deep-sea cores with dates of palaeomagnetic polarity transitions during the Upper Pliocene (1.5–4.3 Myr ago). Dating of the climatic record was obtained by determining the ages of three biostratig-raphic and two palaeoclimatic horizons by reference to palaeomagnetically dated sequences in New Zealand—the core was taken from a site on the west side of the North Island. Any such correlation will be affected by dating errors which can arise from mis-identification of stratigraphic horizons when relating magnetic and climatic data and from non-uniform sedimentation rates when determining relative timescales between horizons. While there is uncertainty in the dating, Doake's results appear to show a significant correlation between climatic change and reversals of the Earth's magnetic field—the chance that the number of observed coincident dates will be random can be as low as 3×10^{-4}. For one of his models there appears to be a probability of about 0.4 that a climatic event will cause a reversal of the magnetic field. However, because of possible dating errors, it is not clear in the association of magnetic events and climatic change which is cause and which is effect.

The Blake event, about 110 000 yr ago, first detected in deep-sea cores by Smith and Foster (1969), has since been reported in a long sediment core from Lake Biwa, Japan. Similar events were observed lower in the core by Kawai *et al* (1972) and estimated to be about 180 000 yr old (the Biwa I event) and about 295 000 yr (the Biwa II event). Yaskawa (1974) constructed an age against depth curve for the Lake Biwa core using ^{14}C age determinations in the upper part and fission track dating of volcanic ash layers between depths of 40 and 100 m. Excursions that are believed to correlate with the Lake Biwa events have also been detected in deep-sea cores (Wollin *et al* 1971b, Yaskawa 1974)—see figure 7.9. The magnetic events coincide with lows in organic carbon in the Lake Biwa core (see figure 7.10) which are believed to be related to periods of cool climate and low productivity of the lake (Kawai *et al* 1975). The Blake event apparently occurred about the same time as the sudden buildup of ice and climatic cooling that immediately followed and ended the last interglacial stage. Studies of the oxygen isotope ratio in deep-sea cores by Shackleton (1976, 1977) and Johnson (1978) and of changes in sea level recorded in tropical coral reef terraces by Matthews (1972) suggest that, about 110 000 to 115 000 yr ago sea level fell 60–70 m in less than 10 000 yr. The rate of growth of ice sheets would have produced a volume of ice of about 28×10^{6} km^{3} (equivalent to the present Antarctic and Greenland ice sheets) in less than 10 000 yr. Climatic changes on timescales of 10^{3}–10^{5} yr may be strongly modulated by changes in the Earth's orbital geometry and their effects on the seasonal distribution of insolation. A climatic cooling at the time of the Blake event would be predicted by the Milankovitch insolation mechanism, although the ice buildup was extremely rapid.

Creer *et al* (1980) carried out palaeomagnetic and palaeontological studies

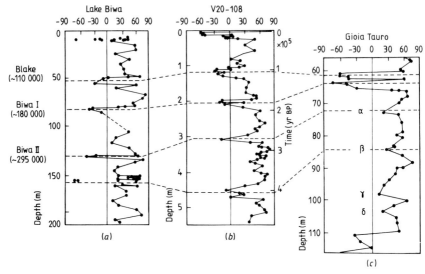

Figure 7.9 (a) Magnetic inclination plotted against depth in the Lake Biwa core: results of preliminary measurements of samples taken at 5 m intervals (Yaskawa 1974). Estimated dates of excursions based on fission-track dating of volcanic ash layers are given at left. (b) Magnetic inclination plotted against depth in deep-sea core V20-108 (Wollin *et al* 1971b). Timescale is based on the assumption of constant sedimentation rates between the top of core (zero age) and the Brunhes–Matuyama boundary. (c) Magnetic inclination plotted against depth for the Gioia Tauro sequence. Major unconformity occurs at about 55 m. Correlation of double event between 60 and 65 m designated as the Blake Event by Creer *et al* (1980), with the Blake Event and Biwa I event, assumes low sedimentation rates or a hiatus in the sequence between the two events. The other correlations are after Creer *et al* (1980) (after Rampino 1981).

on a 250 m core from Gioia Tauro, Italy. They found a number of possible excursions including the Blake event which they suggested was a double event lasting about 50 000 yr (between about 105 000 and 155 000 yr ago). They also attempted to correlate these excursions with similar events reported from Lake Biwa (Yaskawa 1974) and in the deep-sea core V20-108 (Ninkovich *et al* 1966). There are, however, problems with their interpretation of the Gioia Tauro record, the dating of which was based on magnetostratigraphy back to the Matuyama–Gauss transition and assumptions regarding the rate of sedimentation. Although a short double Blake event was reported by Denham (1976) in deep-sea cores, its duration is less than 13 000 yr in the Lake Biwa core and in the original finding in a deep-sea core by Smith and Foster (1969). A further problem is that the Biwa I event was not seen in the Gioia Tauro core. Rampino (1981) has offered an alternative interpretation of the Gioia

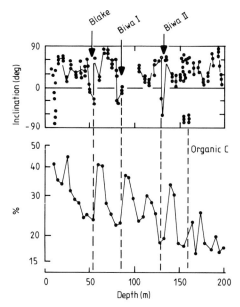

Figure 7.10   Magnetic inclination and percent organic carbon plotted against depth in the Lake Biwa core. Positions of Blake, Biwa I and Biwa II magnetic events are shown by arrows (Kawai 1974) (after Rampino 1979).

Tauro section which avoids these difficulties. He suggested that the long, double Blake event in the Gioia Tauro core is the result of changes in the sedimentation rate and perhaps an undetected hiatus. A change in the nature or rate of sedimentation is suggested by the inferred difference in magnetic mineral content between the interval recording the 'Blake event' and the overlying and underlying sediments. Rampino obtained revised dates for the excursions in the Gioia Tauro core based on those seen in the V20-108 and Lake Biwa cores (see figure 7.9). The revised estimates of the ages of the excursions show an approximate cycle of about 100 000 yr which is similar to that of the variation in the eccentricity of the Earth's orbit (see figure 7.11). The excursions (if real) seem to occur at times of maximum eccentricity giving some support to a possible connection between orbital parameters of the Earth and magnetic fluctuations as discussed earlier (Wollin *et al* 1977, 1978, Rampino 1979).

The Lake Mungo excursion about 30 000 yr ago (see §4.3) is also associated with a time of apparent rapid cooling immediately following a major climatic warming and retreat of ice. This brief interval of warming (less than 5000 yr) is not well recorded in most deep-sea cores and its extent is somewhat controversial. The Biwa I magnetic excursion also appears to have coincided with a time of rapid growth of ice sheets as indicated by oxygen isotope curves

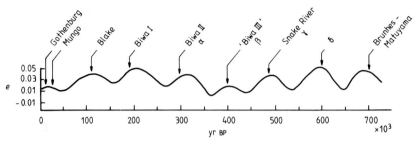

Figure 7.11   Eccentricity (*e*) of the Earth's orbit over the past 725 000 yr (after Kominz *et al* 1979). Arrows indicate inferred ages of magnetic excursions. Age of Snake River event is $(480 \pm 50) \times 10^3$ yr. Age of the δ event is uncertain, as it has only been reported in the Gioia Tauro section (after Rampino 1981).

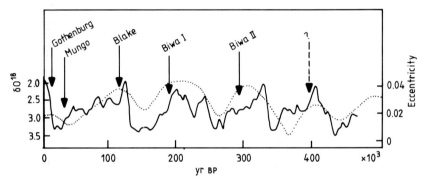

Figure 7.12   Record of oxygen isotope variations during the past 475 000 yr in sub-Antarctic deep-sea cores (Hays *et al* 1976). Low values of $\delta^{18}O$ indicate small ice volumes; high $\delta^{18}O$ values indicate large ice volumes. Times of five possible magnetic excursions are shown by full arrows. (These excursions were not detected in deep-sea cores used to construct record of $\delta^{18}O$.) Time of a sixth magnetic excursion, detected in some deep-sea cores and unnamed at present, is shown by broken arrow. Dotted line is plot of the eccentricity of the Earth's orbit (after Rampino 1979).

of deep-sea sediment cores (see figure 7.12). Furthermore, it took place when the Milankovitch mechanism would suggest growth of global ice sheets. However, the expansion of the ice sheets was again very rapid, as seen in the oxygen isotope curves. (It must be pointed out that these magnetic events have not been detected in the same deep-sea cores that were used for isotopic studies, so that any correlation of climate with magnetic events is indirect.) On the other hand, the time of the Biwa II event was apparently not accompanied by rapid changes in ice volume, as indicated in oxygen isotope studies of most deep-sea cores (see figure 7.12). However, the high deposition rate of the Lake Biwa core (~ 50 cm/1000 yr) indicates a marked decrease in productivity, interpreted as a regional cooling at the time of the Biwa II event, but it must be confessed that the details of climatic changes that occurred about that time are not well known. An as yet unnamed magnetic excursion that has been detected in deep-sea cores and dated at about 400 000 yr also apparently correlates with a time of rapid ice growth and immediately follows a time of major melting of northern hemisphere ice sheets (figure 7.12). Indications of such an event are also found in the Lake Biwa core (figure 7.9). Although not well documented at present, this adds further support to the possible association of climate and geomagnetism.

It must be stressed that coincidence of events, such as changes in global ice volume, magnetic excursions, and short-term rapid glaciations and climatic coolings, do not in themselves prove cause and effect relationships. Also other magnetic excursions that have been reported apparently did not occur at the same times as major ice volume changes and rapid cooling as recorded in oxygen isotope curves of deep-sea cores (Verosub and Banerjee 1977). Again many of the reported excursions are poorly dated, and the reality of several magnetic events has been questioned.

It is worth noting that the last four (and possibly five) times of maximum eccentricity of the Earth's orbit appear to be closely followed by magnetic excursions (see figure 7.11). Eccentricity maxima also appear to coincide with brief periods (around 10^4 yr) of minimum ice cover (the interglacial periods) that followed rapid melting of northern hemisphere ice. Some excursions (for example, the Lake Mungo excursion) are not associated with a maximum of orbital eccentricity, but they do seem to have followed times of rapid ice melting. It appears that if the excursion occurs at an unfavourable time in the Milankovitch insolation cycle (as with the Gothenburg excursion about 13 500 yr ago), then only a brief glacial readvance is observed, but if the geomagnetic trigger occurs at a 'sensitive' phase of the Milankovitch cycle, then a full glacial stage may be initiated.

Williams (1981) claims to have found evidence for cyclic laminations in late Precambrian (~ 680 Myr) glacial varves in the Elatina Formation at Pichi Richi Pass, South Australia. 11, 22, and 90 year cycles were found which he identified with sunspot periods. Longer cycles of 145 and 290 years correlated with solar and climatic periods as evidenced by tree-ring studies. Williams suggested that palaeointensity variations may account for the apparent enhanced solar control of climate in the geological past. The incidence of solar flares and the overall level of flare particle emission vary directly with the 11 year solar cycle. Such corpuscular radiation is normally deflected at the bow shock of the magnetosphere, but some eventually may reach the Earth's atmosphere near the magnetic poles (Lanzerotti 1977). If the dipole component of the geomagnetic field were to weaken or disappear over a few thousand years during a polarity reversal, the shielding effect of the magnetosphere would be much reduced or eliminated. Solar corpuscular radiation could then penetrate the atmosphere to lower levels. The varve cycles of the Elatina Formation therefore may be due ultimately to a weakening of the geomagnetic field during a polarity reversal.

Siscoe *et al* (1976) have estimated the altitude of the magnetopause during reversals of the Earth's magnetic field on the assumption that the present non-dipole field is representative of the minimum field during a reversal. Since, unlike the dipole field, the non-dipole field has a large variation in longitude, Siscoe *et al* calculated the stagnation point of the solar wind as a function of longitude for three latitudes representing the annual latitudinal range of the incident solar wind direction due to the obliquity of the ecliptic. The calculation was carried out for three solar wind stagnation pressures corresponding to the average and extreme pressures measured by the satellite *Explorer 33*. They found that the average position of the stagnation point lies between one and two Earth radii—well above the height where the ionosphere has any important effects on the solar wind. (Sheldon and Kern (1972) estimated that the ionosphere can stop the solar wind at an altitude of approximately 700 km.) Only for extreme solar wind pressures and very restricted regions of longitude and latitude could the solar wind penetrate sufficiently deeply to have any appreciable influence on the lower atmosphere.

An alternative explanation for the varve cycle of the Elatina formation is the observation by Carmichael (1967) that, during the late Precambrian and early Palaeozoic, about 700–400 Myr ago, the geomagnetic field apparently was much weaker than it is today. He estimated a palaeointensity near 10% of the present value at about the time the Elatina Formation was deposited.

7.5 Magnetic reversals and faunal extinctions

Uffen (1963) proposed that if the Earth's magnetic field were reduced to a very low value during a polarity change, the solar wind and a large proportion of cosmic rays would be able to reach the surface of the Earth. He suggested that the biological effects of the increase in radiation might affect the course of evolution and bring about the extinction and transformation of species at the time of a polarity change. Uffen argued on palaeontological grounds that rates of evolution were exceptionally high at times when the Earth's magnetic field was undergoing many changes in polarity. The estimated time interval of about 10^4 yr when the Earth's dipole field is greatly reduced during a reversal is almost instantaneous on the geologic timescale but is very many generations for most living organisms. There are, of course, other causes of mutation, but the exposure to increased radiation would be sudden on the geologic timescale and could contribute to both heritable variation and natural selection. The most striking unexplained fact from palaeontology is that biological evolution has progressed in bursts. Quite adequate explanations have been given for the extinction of whole species by climatic or other environmental changes but the sudden appearance of new forms of life on which the selection pressures of the time would act has not been adequately explained.

A number of investigations (see e.g. Harrison and Funnell 1964, Hays and Opdyke 1967, Watkins and Goodell 1967) have shown that the extinction of certain species of marine organisms occurred at about the time of the most recent reversal of the Earth's magnetic field. However, the reality of any correlation is, in principle, statistical, and there is just not sufficient data. Moreover, Hays *et al* (1969) found that most polarity changes during the last 4 Myr that have extinctions associated with them in Antarctic cores are not the same as those that have extinctions associated with them in equatorial cores from the Pacific. Finally, since only a few organisms become extinct near a polarity change, and some have survived such changes before becoming extinct, it is clear that a polarity change alone is probably insufficient to cause an extinction.

Simpson (1966) attempted to correlate accelerations in the rate of evolution during the Phanerozoic with reversals of the Earth's magnetic field. He compared the change in the percentage of new species as a function of existing species with the occurrence of reversely magnetised rocks. However, as McElhinny (1973) has pointed out, it is the *frequency* of reversals that is the important parameter. McElhinny compared his estimate of the variation in reversal frequency with Newell's (1963) estimate of the rate of organic evolution. He found that accelerations in evolution seemed to occur at times when the reversal rate was either a minimum or a maximum indicating no significant correlation.

A fundamental difference in the polarity history occurred during the latest Cretaceous through the entire Cenozoic, which is represented by a rapidly oscillating reversal sequence. Helsley and Steiner (1974) suggested that major geologic eras, defined on the basis of major changes in the fossil record, may end with a low frequency of reversals and begin with a high reversal frequency. They speculated that the major changes in life forms that take place at these times may have resulted from evolutionary adaptation during the long periods of constant polarity in which the magnetic field is used in the regulation of some necessary or vital function; and then this changes to the disadvantage of some forms. Regardless of the mechanism involved, it is clear that major crises in the Earth's biota at the Palaeozoic–Mesozoic boundary and at the Mesozoic–Cenozoic boundary are associated with major changes in the tempo of polarity changes in the Earth's magnetic field. The faunal extinctions near the end of the Cretaceous and Permo–Carboniferous periods coincide with the close of long intervals of dominantly one magnetic polarity. Kent (1977) has examined the magnetostratigraphy in relation to planktonic foraminiferal changes over an apparently continuous Cretaceous–Tertiary boundary section in the pelagic limestones at Gubbio, northern Italy. He found that this planktonic foraminiferal change and hence the Cretaceous–Tertiary boundary was not coincident with any particular polarity change in the Gubbio section, thus ruling out any simple explanation based on a relationship between faunal change and possible effects on organisms of such a reversal.

Quantitative estimates of the effects on the Earth and atmosphere at a polarity change have been given by Black (1967), Waddington (1967) and Harrison (1968). The surface of the Earth is shielded from cosmic rays by the geomagnetic field and the atmosphere. At the poles, all particles can reach the atmosphere—at the equator, only particles with energies greater than 15 GeV. If the present geomagnetic field were reduced to zero during a polarity transition the increased radiation dose would amount to 10% at the equator and zero at the poles. If the decrease took place over 1000 years, this amounts to an increase in dose of 0.01% per year. The rate of change of dose with sunspot cycle is 0.4% per year, so the possible effects are negligible. Even complete dumping of the energetic particles in the Van Allen radiation belts would not give rise to the necessary increased dosages. A further possibility is that the solar wind might produce ozone in the high atmosphere which would absorb radiation and produce large changes in climate. However, Black (1967) has shown that any ozone produced in this way is but a small fraction of that produced by ultraviolet light. On the other hand, Reid *et al* (1976) have suggested that the reduced magnetic field during a reversal could permit solar proton events seriously to deplete stratospheric ozone. Crutzen *et al* (1975) had shown that a few such intense events a year could materially increase the amount of NO in the stratosphere. Ozone is catalytically destroyed by NO, thus reducing the efficiency of the ozone shield which protects the surface of the Earth from harmful solar ultraviolet radiation.

The only possibility of a causal connection between magnetic reversals and the extinction of certain species seems to be that there is a climatic change at the time of a reversal (see §7.4). There is some evidence (Harrison 1968) that the magnetic field of the Earth has some control over the temperature of the upper atmosphere; also, removal of the Earth's magnetic field would cause large increases in ionisation at certain levels in the upper atmosphere. What climatic changes, if any, would be brought about by such effects is not known. However, a causal connection between magnetic reversals and climatic fluctuations seems unlikely, since the timescale of climatic variations is a good deal shorter than the interval between reversals.

In 1975 Blakemore found that certain aquatic bacteria are magnetotactic and tend to swim along the field lines of the Earth's magnetic field. This is in contrast to most other bacteria which are chemotactic or phototactic. Even when magnetotactic bacteria are killed, they line up along the direction of the Earth's magnetic field. When viewed in a transmission electron microscope, a chain of particles (opaque to the electron beam) is seen more or less parallel to

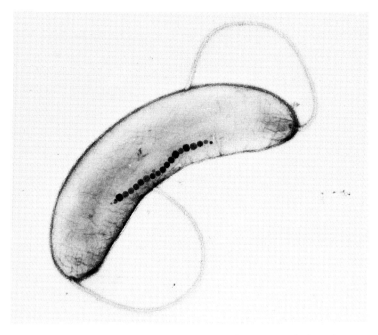

Figure 7.13 Magnetotactic bacterium seen in a transmission electron micrograph. A chain of particles (magnetosomes) can be seen that are opaque to the electron beam (black spots) aligned more or less parallel to the long axis of the cell. The magnification is approximately 28 000 diameters (after Blakemore and Frankel 1981).

the long axis of the cell (see figure 7.13). These particles, called magnetosomes, consist of magnetite (Fe_3O_4), an oxide of iron. It has been shown that single domain particles of magnetotactic bacteria synthesise magnetosomes and carry them with them. Matsuda *et al* (1983) have studied the morphology and structure of magnetosomes using high resolution electron microscopy. They found them to be single crystals with a hexagonal prism shape truncated by {111} planes. They are believed to consist of magnetite, since the spacings and orientations of all the observed fringes agree with those of magnetite. Their polarity can be changed by exposure to a brief but strong magnetic field in the opposite direction to the ambient field. Their original polarity can be restored by a second exposure to a magnetic field antiparallel to the first. Presumably, during a reversal of the Earth's magnetic field, the polarity of the magnetotactic bacteria in both hemispheres becomes reversed as well.

Laboratory experiments have shown that it is the vertical component of the geomagnetic field that is the relevant factor in determining the polarity of bacterial populations in natural environments. In the northern hemisphere most magnetotactic bacteria are north seeking, i.e. they tend to move downwards, keeping to the sediments and away from surface water. Since magnetotactic bacteria are anaerobic or microaerophilic, a tendency to migrate downwards would help to avoid the toxic effects of the greater concentration of oxygen in surface waters.

Blakemore *et al* later (1980) found several types of magnetotactic bacteria in sediments of the southern hemisphere. They were consistently south seeking, i.e. they moved downwards like those in the Northern hemisphere. Electron microscopy showed that they contained internal chains of electron-opaque particles similar to those observed in magnetotactic bacteria from the northern hemisphere. Like the northern hemisphere bacteria, their magnetic polarity can be permanently reversed in the laboratory and they cannot be demagnetised. Frankel *et al* (1981) found approximately equal numbers of north seeking and south seeking bacteria in freshwater and marine sediments at Fortaliza, Brazil, close to the geomagnetic equator. This confirms the hypothesis that the vertical component of the geomagnetic field determines the predominant polarity of magnetotactic bacteria. Because of the horizontal orientation of the magnetic field at the geomagnetic equator, the motion of magnetotactic bacteria there will be directed horizontally. This would be advantageous to bacteria of either polarity in reducing detrimental upward migration as compared to random motion.

If the geomagnetic field is reversed, magnetotactic bacteria would be directed away from their natural environment into one in which they might be unable to survive. This seems to give a wholly convincing explanation of the extinction phenomenon. It is all the more convincing because it is an effect against which a species cannot protect itself on the basis of past experience. Presumably the mechanism operates within some range of *intensity* of the magnetic field, but if it takes its *direction* from the field it is incapable of knowing when that direction has changed relative to anything else.

All this seems reasonable if there is a sufficiently sudden change in the direction of the field. Whether a species has any capability of adjusting to a gradual change is not known. It does not follow, of course, that the mode of extinction of the smallest living organisms has any connection with that of larger creatures. However, since 1975, magnetite has been found in many other organisms besides bacteria—chitons (a class of marine molluscs), honeybees, butterflies, homing pigeons, dolphins, tuna fish and sea turtles.

It is noteworthy that the two great extinctions about 225 and 65 Myr ago occurred in the mixed magnetic intervals 230–204 Myr and 70–50 Myr ago. During each of these there were many polarity reversals, any one of which could have been catastrophic for any species for which some crucial instinctive behaviour was linked to the geomagnetic field. The former of these two mixed intervals followed an interval of about 25 Myr during which no reversals at all have been recorded; the latter followed an interval of about 40 Myr which was interrupted by only one relatively short-lived reversal. In both cases the biosphere had had time to evolve as though polarity never changed before a sequence of 'rapid' changes began.

References

Amerigian C 1974 Sea-floor dynamic processes as the possible cause of correlations between palaeoclimatic and palaeomagnetic indices in deep-sea sedimentary cores *Earth Planet. Sci. Lett.* **21** 321

Berger A 1976 Obliquity and precession for the last 5 000 000 years *Astron. Astrophys.* **51** 127

——1977a Support for the astronomical theory of climatic change *Nature* **269** 44

——1977b Long-term variations of the Earth's orbital elements *Celestial Mech.* **15** 53

——1978 Long-term variations of caloric insolation resulting from the Earth's orbital elements *Quat. Res.* **9** 139

Black D I 1967 Cosmic-ray effects and faunal extinctions at geomagnetic field reversals *Earth Planet. Sci. Lett.* **3** 225

Blakemore R 1975 Magnetotactic bacteria *Science* **190** 377

Blakemore R P and Frankel R B 1981 Magnetic navigation in bacteria *Sci. Am.* **245** (12) 42

Blakemore R P, Frankel R B and Kalmijn Ad J 1980 South-seeking magnetotactic bacteria in the southern hemisphere *Nature* **286** 384

Bonatti E and Gartner S 1973 Caribbean climate during Pleistocene ice ages *Nature* **244** 563

Bucha V 1976a Variations of the geomagnetic field, the climate and weather *Stud. Geophys. Geod.* **20** 149

——1976b Changes in the geomagnetic field and solar wind—causes of changes of climate and atmospheric circulation *Stud. Geophys. Geod.* **20** 346

——1977 The mechanism of solar terrestrial relationships and changes of the atmospheric circulation *Stud. Geophys. Geod.* **21** 3

——1980 Mechanism of the relations between the changes of the geomagnetic field, solar corpuscular radiation, atmospheric circulation and climate *J. Geomag. Geoelec.* **32** 217

Bullard E C 1968 Reversals of the Earth's magnetic field *Phil. Trans. R. Soc.* A **263** 481

Carmichael C M 1967 An outline of the intensity of the palaeomagnetic field of the Earth *Earth Planet. Sci. Lett.* **3** 351

Chappel J 1975 On possible relationships between Quaternary glaciation, geomagnetism and vulcanism *Earth Planet. Sci. Lett.* **26** 370

Chave A D and Denham C R 1979 Climatic changes, magnetic intensity variations and fluctuations of the eccentricity of the Earth's orbit during the past 2 000 000 years and a mechanism which may be responsible for the relationship—a discussion *Earth Planet. Sci. Lett.* **44** 150

Chiu Y T 1974 Archaeomagnetism and archaeoclimatic 'forecast'? *Nature* **250** 642

Courtillot V, Ducruix J and Le Mouel J-L 1978 Sur une accélération récente de la variation séculaire du champ magnétique terrestre *C.R. Acad. Sci., Paris* D **287** 1095

Courtillot V, Le Mouel J-L, Ducruix J and Cazenave A 1982 Geomagnetic secular variation as a precursor of climatic change *Nature* **297** 386

Creer K M, Readman P W and Jacobs A M 1980 Palaeomagnetic and palaeontological dating of a section at Gioia Tauro, Italy: identification of the Blake event *Earth Planet. Sci. Lett.* **50** 289

Crutzen P J, Isaksen I S A and Reid G C 1975 Solar proton events: stratospheric sources of nitric oxide *Science* **189** 457

Denham C R 1976 Blake polarity episode in two cores from the Greater Antilles outer ridge *Earth Planet. Sci. Lett.* **29** 422

Doake C S M 1977 A possible effect of ice ages on the Earth's magnetic field *Nature* **267** 415

——1978 Climatic change and geomagnetic field reversals: a statistical correlation *Earth Planet. Sci. Lett.* **38** 313

Eddy J A, Gilliland R L and Hoyt D V 1982 Changes in the solar constant and climatic effects *Nature* **300** 689

Ericson D B and Wollin G 1956 Micropalaeontological and isotopic determinations of Pleistocene climates *Micropalaeontology* **2** 257

Evans D L and Freeland H J 1977 Variations in the Earth's orbit: Pacemaker of the Ice ages? *Science* **198** 528

Fairbridge R W 1962 New radiocarbon dates of Nile sediments *Nature* **196** 108

——1977 Global climate change during the 13 500 BP Gothenburg geomagnetic excursion *Nature* **265** 430

Frankel R B, Blakemore R P, Torres de Araujo F F, Esquivel M S and Danon J 1981 Magnetotactic bacteria at the geomagnetic equator *Science* **212** 1269

Glass B P and Heezen B C 1967 Tektites and geomagnetic reversals *Sci. Am.* **217**(7) 32

Goodell H G and Watkins N D 1968 The palaeomagnetic stratigraphy of the Southern Ocean: 20°W to 160°E longitude *Deep-sea Res.* **15** 89

Goodell H G, Watkins N D, Mather T T and Koster S 1968 The Antarctic glacial history recorded in sediments of the Southern Ocean *Palaeogeogr., Palaeoclimatol., Palaeoecol.* **5** 41

Harrison C G A 1968 Evolutionary processes and reversals of the Earth's magnetic field *Nature* **217** 46

——1974 The palaeomagnetic record from deep-sea sediment cores *Earth Sci. Rev.* **10** 1

Harrison C G A and Funnel B M 1964 Relationship of palaeomagnetic reversals and micropalaeontology in two late Cenozoic cores from the Pacific Ocean *Nature* **204** 566

Harrison C G A and Prospero J M 1974 Reversals of the Earth's magnetic field and climatic changes *Nature* **250** 563

Hays J D 1967 Quaternary sediments of the Antarctic Ocean *The Quaternary History of the Ocean Basins, Oceanography* **4** 117

Hays J D, Imbrie, J and Shackleton N J 1976 Variations in the Earth's orbit: Pacemaker of the Ice ages? *Science* **194** 1121

——1977 Variations in the Earth's orbit: Pacemaker of the Ice ages? *Science* **198** 529

Hays J D and Opdyke N D 1967 Antarctic radiolaria, magnetic reversals and climatic change *Science* **158** 1001

Hays J D, Saito T, Opdyke N D and Burckle L 1969 Pliocene/Pleistocene sediments of the Equatorial Pacific; their palaeomagnetic, biostratigraphic and climatic record *Geol. Soc. Am. Bull.* **80** 1481

Hecht A D 1977 (Reply by Fairbridge R W) Geomagnetism and climate *Nature* **268** 669

Heirtzler J R 1970 The palaeomagnetic field as inferred from marine magnetic studies *J. Geomag. Geoelec.* **22** 197

Helsley C E and Steiner M B 1974 Palaeomagnetism of the lower Triassic Moenkopi Formation *Bull. Geol. Soc. Am.* **85** 457

Hines C O and Halevy I 1977 On the reality of a certain sun-weather correlation *J. Atmos. Sci.* **34** 382

Imbrie J and Imbrie J Z 1980 Modelling the climatic response to orbital variations *Science* **207** 943

Johnson R G 1978 Initial glacial eustatic sea-level fall in the early Wisconsin calculated from temperature corrected isotope ratios in cores *Geol. Soc. Am. Abstr.* **10** 429

Kawai N 1974 Restrained photosynthesis during Brunhes field transition *Palaeolimnology of Lake Biwa and the Japanese Pleistocene* vol 2, ed S. Horie (Otsu: Kyoto University) p 59

Kawai N, Yaskawa K, Nakajima T, Torii M and Horie S 1972 Oscillating geomagnetic field with a recurring reversal discovered from Lake Biwa *Proc. Jap. Acad.* **48** 186

Kawai N, Yaskawa K, Nakajima T, Torii M and Natsuhara N 1975 Voices of geomagnetism from Lake Biwa *Palaeolimnology of Lake Biwa and the Japanese Pleistocene* vol 3, ed S. Horie (Otsu: Kyoto University) p 143

Kennett J P and Watkins N D 1970 Geomagnetic polarity change, volcanic maxima and faunal extinction in the South Pacific *Nature* **227** 930

Kennett J P, Watkins N D and Vella P 1971 Palaeomagnetic chronology of Pliocene–Early Pleistocene climates and the Plio–Pleistocene boundary in New Zealand *Science* **171** 276

Kent D V 1977 An estimate of the duration of the faunal change at the Cretaceous–Tertiary boundary *Geology* **5** 769

——1982 Apparent correlation of palaeomagnetic intensity and climatic records in deep-sea sediments *Nature* **299** 538

——1983 Geomagnetic excursions and climate change *Nature* **302** 455

Kent D V and Opdyke N D 1977 Palaeomagnetic field intensity variation recorded in a Brunhes epoch deep-sea sediment core *Nature* **266** 156

King J W 1974a Weather and the Earth's magnetic field *Nature* **247** 131

——1974b Geomagnetism and the tropospheric circulation *Nature* **252** 370

——1975 Sun–weather relationships *Astronaut. Aeronaut.* **13** 10

Knight J W and Sturrock P A 1976 Solar activity, geomagnetic field, and terrestrial weather *Nature* **264** 239

Kominz M A, Heath G R, Ku T-L and Pisias N-G 1979 Brunhes timescales and the interpretation of climatic change *Earth Planet. Sci. Lett.* **45** 394

Kominz M A and Pisias N G 1979 Pleistocene climate: Deterministic or stochastic *Science* **204** 171

Kukla G, Berger A, Lotti R and Brown J 1981 Orbital signature of interglacials *Nature* **290** 295

Lambeck K and Cazenave A 1976 Long-term variations in the length of day and climatic change *Geophys. J.* **46** 555

Lanzerotti L J 1977 *The Solar output and its variation* ed O R White (Boulder: Colorado Assoc. Univ. Press) p 383

Le Mouel J-L, Madden T R, Ducruix J and Courtillot V 1981 Decade fluctuations in geomagnetic westward drift and Earth rotation *Nature* **290** 763

McDougall I and Wensink H 1966 Palaeomagnetism and geochronology of the Pliocene–Pleistocene lavas in Iceland *Earth Planet. Sci. Lett.* **1** 232

McElhinny M W 1973 *Palaeomagnetism and plate tectonics* (Cambridge: Cambridge University Press)

Mankinen E A and Dalrymple G B 1979 Revised geomagnetic polarity timescale for the interval 0–5 Myr BP *J. Geophys. Res.* **84** 615

Margolis S V and Herman Y 1980 Northern hemisphere sea-ice and glacial development in the late Cenozoic *Nature* **286** 145

Matsuda T, Endo J, Osakabe N and Tonomura A 1983 Morphology and structure of biogenic magnetite particles *Nature* **302** 411

Matthews R K 1972 Dynamics of the ocean cryosphere system: Barbados data *Quat. Res.* **2** 368

Milankovitch M 1941 *K. Serb. Akad. Beogr. Spec. Publ.* **132** (Translated—Israel program for Scientific Translation, Jerusalem 1969)

Moore T C, Pisias N G and Dunn D A 1982 Carbonate time series of the Quaternary and late Miocene sediments in the Pacific ocean; a spectral comparison *Marine Geol.* **46** 217

Morner N-A and Lanser J 1975 Palaeomagnetism in deep-sea core A179-15 *Earth Planet. Sci. Lett.* **26** 121

Newell N D 1963 Crises in the history of life *Sci. Am.* **208** (2) 76

Ninkovich D, Opdyke N, Heezen B C and Foster J H 1966 Palaeomagnetic stratigraphy and tephrochronology in North Pacific deep-sea sediments *Earth Planet. Sci. Lett.* **1** 476

Oerlemans J 1980 Model experiments on the 100 000 yr glacial cycle *Nature* **287** 430

Opdyke N D, Glass B, Hays J D and Foster J 1966 Palaeomagnetic study of Antarctic deep-sea cores *Science* **154** 349

Pittock A B 1978 A critical look at long-term sun–weather relationships *Rev. Geophys. Space Phys.* **16** 400

Rampino M R 1979 Possible relationships between changes in global volume, geomagnetic excursions and the eccentricity of the Earth's orbit *Geology* **7** 584

——1981 Revised age estimates of Brunhes palaeomagnetic events: support for a link between geomagnetism and eccentricity *Geophys. Res. Lett.* **8** 1047

Reid G C, Isaksen I S A, Holzer T E and Crutzen P J 1976 Influence of ancient solar-proton events on the evolution of life *Nature* **259** 177

Roberts W O and Olson R H 1973a Geomagnetic storms and wintertime 300 mb trough development in the North Pacific–North America area *J. Atmos. Sci.* **30** 135

——1973b New evidence for effects of variable solar corpuscular emission on the weather *Rev. Geophys. Space Phys.* **11** 731

Rooth C G H, Emiliani C and Poor H W 1978 Climate response to astronomical forcing *Earth Planet. Sci. Lett.* **41** 387

Ruddiman W F and McIntyre A 1976 Northeast Atlantic palaeoclimatic changes over the past 600 000 years *Geol. Soc. Am. Mem.* **145** 111

Sawyer J S 1974 Geomagnetism and the tropospheric circulation *Nature* **252** 368

Shackleton N J 1976 Oxygen-isotope evidence relating to the end of the last interglacial at the substage 5e to 5d transition about 115 000 years ago *Geol. Soc. Am. Abst.* **8** 1099

——1977 Carbon 13 in Uvigerina: Tropical rainforest history and the Equatorial Pacific carbonate dissolution cycles *The fate of fossil fuel CO_2 in the oceans* ed N R Anderson and A Malahoff (New York: Plenum) p 401

Shackleton N J and Opdyke N D 1973 Oxygen isotope and palaeomagnetic stratigraphy of equatorial Pacific core V28-238: oxygen isotope temperature and ice volume on a 10^4 and 10^5 year scale *Quat. Res.* **3** 39

——1976 Oxygen isotope and palaeomagnetic stratigraphy of Pacific core V28-239, Late Pliocene to Latest Pleistocene *Geol. Soc. Am. Mem.* **145** 449

Shapley A H, Kroehl H W and Allen J H 1975 Solar terrestrial physics and meteorology: a working document *Spec. Comm. Solar Terr. Phys. Rep.* (Washington DC: Nat. Acad. Sci.)

Sheldon W R and Kern J W 1972 Atmospheric helium and geomagnetic field reversals *J. Geophys. Res.* **77** 6194

Simpson J F 1966 Evolutionary pulsation and geomagnetic polarity *Geol. Soc. Am. Bull.* **77** 197

Siscoe G L, Chen C K and Harel M 1976 On the altitude of the magnetopause during geomagnetic reversals *J. Atmos. Terr. Phys.* **38** 1327

Smith J D and Foster J H 1969 Geomagnetic reversal in Brunhes normal polarity epoch *Science* **163** 565

Sternberg R S and Damon P E 1979 Re-evaluation of possible historical relationship between magnetic intensity and climate *Nature* **278** 36

Stolov H L and Shapiro R 1971 Report on an investigation of solar corpuscular influences on the general circulation *Symp. solar corpuscular effects in the troposphere and stratosphere* (Moscow: IUGG)

——1974 Investigation of the responses of the general circulation at 700 mbar to solar geomagnetic disturbance *J. Geophys. Res.* **79** 2161

Suarez M J and Held I M 1976 Modelling climatic response to orbital parameter variations *Nature* **263** 46

Uffen R J 1963 Influence of the Earth's core on the origin and evolution of life *Nature* **198** 143

Van Woerkom A J J 1953 The astronomical theory of climate changes *Climatic Change* ed H Shapley (Harvard: Harvard University Press)

Vernekar A D 1972 *Meteorol. Monog.* **12**

Verosub K L and Banerjee S K 1977 Geomagnetic excursions and their palaeomagnetic record *Rev. Geophys. Space Phys.* **15** 145

Waddington C J 1967 Palaeomagnetic field reversals and cosmic radiation *Science* **158** 913

Watkins N D and Goodell H G 1967 Geomagnetic polarity change and faunal extinction in the Southern Ocean *Science* **156** 1083

Weertman J 1961 Stability of ice age ice sheets *J. Geophys. Res.* **66** 3783

——1976 Milankovitch solar radiation variations and ice age ice sheet sizes *Nature* **261** 17

Wigley T M L 1976 Spectral analysis and the astronomical theory of climatic change *Nature* **264** 629

Wilcox J M 1975 Solar activity and the weather *J. Atmos. Terr. Phys.* **37** 237

Williams G E 1981 Sunspot periods in the late Precambrian glacial climate and solar–planetary relations *Nature* **291** 624

Wollin G, Ericson D B and Ryan W B F 1971a Variations in magnetic intensity and climatic changes *Nature* **232** 549

Wollin G, Ericson D B, Ryan W B F and Foster J H 1971b Magnetism of the Earth and climatic changes *Earth Planet. Sci. Lett.* **12** 175

Wollin G, Kukla G J, Ericson D B, Ryan W B F and Wollin J 1973 Magnetic intensity and climatic changes 1925–1970 *Nature* **242** 34

Wollin G, Ryan W B F, Ericson D B and Foster J H 1977 Palaeoclimate, palaeomagnetism and the eccentricity of the Earth's orbit *Geophys. Res. Lett.* **4** 267

Wollin G, Ryan W B F and Ericson D B 1978 Climatic changes, magnetic intensity variations and fluctuations of the eccentricity of the Earth's orbit during the past

2 000 000 years and a mechanism which may be responsible for the relationship *Earth Planet. Sci. Lett.* **41** 395

Yaskawa K 1974 Reversals, excursions and secular variations of the geomagnetic field in the Brunhes normal polarity epoch *Palaeolimnology of Lake Biwa and the Japanese Pleistocene* vol 2, ed S Horie (Otsu: Kyoto University) p 77

Subject index

virtual dipole moment (VDM) 44, 55,
  60, 90, 98, 105, 148–9
virtual geomagnetic pole (VGP) 42, 53,
  58, 60–9, 86, 94, 98, 101, 103,
  108–9, 140, 148–9, 166–7
viscous remanent magnetism (VRM)
  26–7

volcanism 185

westward drift 9–10, 60, 91–4, 198
Wrangell Mountains, Alaska 107

Yarraman Creek, Australia 33
Yolida Clay, Denmark 102–3

Name index

Abdel-Monem A 165
Abrahamsen N 98, 102–3, 110
Ade-Hall J 7, 33, 35, 102, 150
Aitken M J 89–91
Akimoto S 30, 32
Aldridge K D 141
Allan D W 119–20
Allen J H 199
Alvarez W 77–8, 80, 146, 171–4, 177
Amerigian C 194
Anderson R F 51, 103
Anderson T W 93, 101
Arthur M A 171, 177
Aumento F 102
Avery O E 76
Aziz-ur-Rahman 161, 164

Baag C G 51
Babkine J 87, 106
Bacon M P 51, 103
Balsey J R 35
Banerjee S K 22, 87, 91, 97, 101, 108, 207
Barbetti M F 56–7, 86, 89–92, 97–8
Bartels J 11
Barton C E 94
Berger A 186, 190–2
Berggren W A 166, 174–5
Berglund B 88, 91, 97–8
Bhimasankaram V L S 33
Bingham D K 50, 52, 107–8
Black D 210
Blakely R J 43, 61, 73–4, 141–5, 147, 164, 170

Blakemore R 210–12
Beil U 33
Blow R A 110
Bochev A 126
Bogue S W 67
Bol'shakov A S 52
Bolt B A 136
Bonatti E 195
Bonhommet N 87–90, 106
Briden J C 165–6
Brock A 134, 164
Broecker W C 102
Brown F H 164
Brown J 192
Bucha V 198–9
Buddington A F 35
Bullard E C ix, 11, 117–19, 140, 184
Burakov K S 65
Burckle L H 160, 166, 171, 209
Burek P J 34, 172
Busse F H 15, 18
Butler R F 61, 170–2, 174–5, 179

Cande S C 77, 79–80, 146–7, 171–5
Carmichael C M 31, 208
Cazenave A 199–200
Chamalaun F H 105, 164
Champion D E 104, 106
Chandrasekhar S 124
Channel J E T 27, 171–2
Chappel J 185
Chave A D 195
Cheli P 172–3
Chen C K 208